INNER PLANETS

Greenwood Guides to the Universe
Timothy F. Slater and Lauren V. Jones, Series Editors

Astronomy and Culture
Edith W. Hetherington and Norriss S. Hetherington

The Sun
David Alexander

Inner Planets
Jennifer A. Grier and Andrew S. Rivkin

Outer Planets
Glenn F. Chaple

Asteroids, Comets, and Dwarf Planets
Andrew S. Rivkin

Stars and Galaxies
Lauren V. Jones

Cosmology and the Evolution of the Universe
Martin Ratcliffe

INNER PLANETS

Jennifer A. Grier and
Andrew S. Rivkin

Greenwood Guides to the Universe
Timothy F. Slater and Lauren V. Jones, Series Editors

GREENWOOD PRESS
An Imprint of ABC-CLIO, LLC

A B C CLIO

Santa Barbara, California • Denver, Colorado • Oxford, England

Library of Congress Cataloging-in-Publication Data

Grier, Jennifer A.
Inner planets / Jennifer A. Grier and Andrew S. Rivkin.
p. cm. — (Greenwood guides to the universe)
Includes bibliographical references and index.
ISBN 978–0–313–34430–5 (hardcover : alk. paper) — ISBN 978–0–313–34431–2 (ebook)
1. Inner planets—Popular works. I. Rivkin, Andrew S. II. Title.
QB602.G75 2010
523.4—dc22 2009042491

14 13 12 11 10 1 2 3 4 5

This book is also available on the World Wide Web as an eBook.
Visit www.abc-clio.com for details.

ABC-CLIO, LLC
130 Cremona Drive, P.O. Box 1911
Santa Barbara, California 93116-1911

This book is printed on acid-free paper ∞

Manufactured in the United States of America

Contents

Series Foreword

Not since the 1960s and the Apollo space program has the subject of astronomy so readily captured our interest and imagination. In just the past few decades, a constellation of space telescopes, including the Hubble Space Telescope, have peered deep into the farthest reaches of the universe and discovered supermassive black holes residing in the center of galaxies. Giant telescopes on Earth's highest mountaintops have spied planet-like objects larger than Pluto lurking at the very edges of our solar system and have carefully measured the expansion rate of our universe. Meteorites with bacteria-like fossil structures have spurred repeated missions to Mars with the ultimate goal of sending humans to the red planet. Astronomers have recently discovered hundreds more planets beyond our solar system than within it. Such discoveries give us a reason for capturing what we now understand about the cosmos in these volumes, even as we prepare to peer deeper into the universe's secrets.

As a discipline, astronomy covers a range of topics, stretching from the central core of our own planet outward past the Sun and nearby stars to the most distant galaxies of our universe. As such, this set of volumes systematically covers all the major structures and unifying themes of our evolving universe. Each volume comprises a narrative discussion highlighting the most important ideas of major celestial objects and how astronomers have come to understand their nature and evolution. In addition to describing astronomers' most current investigations, many volumes include perspectives on the historical and premodern understandings that have motivated us to pursue deeper knowledge.

The ideas presented in these assembled volumes have been meticulously researched and carefully written by experts to provide readers with the most scientifically accurate information that is currently available. There are some astronomical phenomena that we just do not understand very well, and the authors have tried to distinguish between which theories have wide consensus and which are still as yet unconfirmed. As astronomy is a rapidly advancing science, it is almost certain that some of the concepts presented in these pages will become obsolete as advances in technology yield previously unknown information. Astronomers share and value a worldview in which our scientific knowledge is subject to change as the scientific enterprise makes new and better observations of our universe. Our understanding of the cosmos evolves over time, just as the universe evolves, and what we learn tomorrow depends on the insightful efforts of dedicated scientists from yesterday and today. We hope that these volumes reflect the deep respect we have for the scholars who have worked, are working, and will work diligently in the public service to uncovering the secrets of the universe.

Lauren V. Jones, Ph.D.
Timothy F. Slater, Ph.D.
University of Wyoming
Series Editors

Preface

THE PLANETS ENTHRALL US

Young children seem to have two natural scientific loves—astronomy and dinosaurs. Walk through your local science museum and see the places where the youngest visitors appear most engaged. Listen to the topics about which they ask the most questions. Both represent mysteries of the past, are surrounded by fantastic stories, and have bearing on the life and fate of humans, now and in the future. Our interest in the night sky is by far the older of the two, since the certain knowledge that our world was home to dinosaurs is a relatively recent event.

The sky was an obsession to all peoples; it provided a source of inspiration and religious focus, a means for determining direction and navigation, times to sow and times to harvest, and much more. Ancient civilizations, as well as more recent ones like ourselves, built special structures to act as observatories of the heavens. The times of sunrise and sunset, the phases of the Moon, and the movements of those particularly interesting stars known as the planets were all recorded and studied from these observatories. The planets, a word that means *wanderers*, were the star-like objects that moved. The rest of the stars were fixed into permanent patterns. The planets were given special significance because of this, and many religions associated their deities with them.

Our obsession continues today. News events centered on missions to the planets are commonplace, as the media seeks to fill the public's deep interest in the latest discoveries. The lessons we have learned have helped us assess the threat of climate change and consider whether we are alone in the solar system. We have sent our machines to fly by the scarred landscape of Mercury, to land on the baked plains of Venus, to drive across the deserts of Mars. We have sent our flesh and blood to visit the Sea of Tranquility. And we, all of us, explore our home planet on a daily basis.

GOALS AND AUDIENCE FOR THIS VOLUME

Education about astronomy, both formally in school and informally because of general interest, is an important endeavor, and continually needs resources that reflect our current understanding. Such resources, like this book, are built on the principles of good education, as well as accurate science. Gone are the days when memorizing a list of facts was considered learning science; today the goal is to learn the processes and understand how scientists think, so that anyone can be a scientist with the situations in their daily lives.

People want to know what's out there, how it got there, and how that impacts their own ideas of their place in it all. And yet educationally the public at large is not well equipped to understand what they read. It can be perceived as a great disservice to all citizens that education in the United States has not placed the necessary emphasis on general science that is required to make people truly literate in our increasingly scientific world. People continue to be taken in by scams, hoaxes, urban legends, and conspiracy theories that could be avoided if they had the opportunity to learn the kind of critical thinking skills inherent in the sciences. Planetary science is in the interesting position of being of high public interest, but low in the public's true ability to understand because educational background is often lacking. What is global warming? What really is the risk of living near a nuclear power plant? Will a solar flare disrupt my cell phone communication? What causes the phases of the Moon (most people would say clouds, by the way)? Is there life on other planets? These common questions are all related to the tools, processes, and topics of planetary science. Understanding how to answer them is not magic, and anyone can become conversant enough with the basics to feel more at home with the answers, and especially with the questions we are still trying to answer. One of the goals of *Inner Planets*, indeed for the all the volumes in the Greenwood Guides to the Universe Series, is to be a resource that will encourage readers to make their own observations, collect their own evidence, demonstrate healthy skepticism, and take time for considered thought related to any topic that they encounter in the media about our place in space and time.

A host of texts at the undergraduate level focus on planetary science, and particularly the inner planets. The authors of this volume did not intend to simply reproduce another general survey of the inner planets, although much of the standard information about the planets is indeed presented here. The purpose of this volume is to provide the unique insight of the authors, both of whom are planetary scientists and educators, into the fundamental content- and process-related underpinnings of the discipline. No text can provide a comprehensive view of the planets, the subject is too broad, and changes too rapidly. Every new space mission to the planets, and every new advance in instrumentation, provides a new flood of data. Instead, *Inner Planets* goes into the details of the basic processes that form and evolve planets, and then how those processes act together to create the complex and dynamic systems found on all worlds in the solar system. A student who becomes familiar with the general processes explained in this volume will develop a deeper understanding of how professional planetary scientists approach their research, what motivates scientists to choose one theory or model over another, how hypotheses are tested and how they change over time, and how anyone can develop a considered scientific intuition for the relative importance of various factors in the creation and evolution of planets. The authors have also tried to present this information in an engaging fashion—so many texts about the subject are dry and packed with factual information that is often better looked up than memorized. An important goal for this volume is for any reader to come away with an appreciation for how amazing the inner planets are, and to find the study of them to be exciting and interesting.

Inner Planets will be useful in a wide range of educational settings, as well as for informal learning. Advanced high school students will find useful background material and an introduction to more sophisticated ideas. They, along with majors and nonmajors at the undergraduate level, will be able to look at some issues surrounding the study of the inner planets in more depth, as well as gain a better appreciation for the nature of error and estimation in science in general. Graduate students in geology, astronomy, and planetary science can use this volume as a solid background text to the inner planets, and also make direct use of the more quantitative ideas, such as atmospheric scale height, saltation particle size, and radiometric decay. *Inner Planets*, along with the rest of the volumes in the Greenwood Guides to the Universe Series, will provide an excellent resource for students and those generally interested in a deeper understanding of astronomical phenomena.

INSIDE THIS BOOK

The authors have chosen to highlight those processes and systems that seem both the most basic, and the most critical, to an understanding of the inner planets. Other writers may well have chosen to emphasize different aspects of planetary science and of the planets themselves, but the authors of this volume have selected those aspects that appear to be of universal use and interest to scientists active in the discipline. They have also focused on those areas that cover the important terms (i.e., jargon) of the field so that any reader will become conversant in the basic language used by planetary scientists. **Large tables of data, sometimes found in other books on the subject, are not presented. Factual information such as size, gravity, and such, have become so easily available on the internet and from other sources that best use of space did not suggest their presentation here.**

Inner Planets is organized into two general sections. The first chapters discuss the major processes related to planetary formation and evolution, e.g., impact cratering, volcanism, radiometric dating, tectonics, atmospheres, magnetospheres, and the role of water. Erosion, often considered separately as a process, is here presented within those chapters that are related, e.g., water erosion in the chapter about hydrology and mass wasting in the chapter about tectonics. Each chapter includes an introduction to the subject, and then a breakdown of the major topics. Images or diagrams have been chosen either to illustrate major concepts, or to show the more interesting and unusual aspects of the topic. Sidebars present a brief look at historical and cultural topics of particular note.

Planetary science changes rapidly. The numbers printed in this book, such as "about 170 giant volcanoes on Venus" might change tomorrow. Where possible, exact numbers are not stated, but instead general ranges or estimates are given that support a given point or provide certain evidence. Where a more definite idea of scope is needed, then specific numbers are provided. Note, however, that it will always be prudent to double check these with the very latest data, since new phenomena are being discovered and redefined constantly.

CONCLUSION

The authors hope that any reader will come away from this book with an appreciation both for how much we have learned about the inner planets, and how many major questions are still open and debated, waiting to be answered. Ideally, *Inner Planets* and the other volumes in the series will provide one more opportunity for people to continue to learn about astronomy for themselves, and to enable them to educate others as teachers, scientists, parents, or interested citizens.

The space program of the United States, while it has had its ups and downs as has any institution, remains a source of national pride. There is a collective understanding that continually pushing the bounds of our knowledge and our understanding of our environment is a sign of a healthy, growing society. The moment we no longer feel the need to look outward, but only inward or down, we will have come to a sad turning point. Hopefully, it will never come to pass. Continuing to ensure that future generations can share meaningfully in the wonder of the planets is not only a joy but also an obligation.

Acknowledgments

We could not have written this book without the help, support, and input of several groups and individuals.

We are pleased to acknowledge and thank the team at Greenwood Press (now ABC-CLIO): Lauren Jones, Kevin Downing, and John Wagner. We greatly appreciate their limitless patience, comments, and encouragement.

We would also like to thank Tim Spahr, who suggested us as possible authors and put us in touch with the Greenwood Press team. Additional thanks go to friends and colleagues who supported us during the writing of this volume, especially Amy Grier.

The content of this book was strongly motivated by exceptionally high-quality instruction and mentorship from Jay Melosh more than 15 years previous. He provided us with insight and appreciation for the essentials of planetary science, both theoretical and experimental.

1

Introduction

A UNIVERSE OF SYSTEMS

The universe is really a series of systems, all connected and interacting. From the gravity of massive super groups of galaxies down to the quantum states of subatomic particles, all phenomena are connected and all are interdependent. The same forces and processes are at work here and across the universe, following the same laws of physics and described by the same mathematical equations. The understanding of one part of our cosmos usually requires at least a general understanding of what the major processes and forces are in the other parts as well, and then, how they are related to one another.

Star Systems

Nestled into one arm of the spiral-shaped Milky Way galaxy is an average-sized yellow-green star. It would at first seem to be quite ordinary, although one might notice that it had no other stellar companion—most stars are formed in pairs or more. But then one would find the teeming and diverse life on the third planet, and would definitely think this stellar system was worth another look.

Star systems (often all referred to as "solar systems" although "Sol" is the name of our particular star) are specifically called systems for a reason. Stars are born from vast collections of **gas** and dust compressed either by gravity or perhaps by the shock waves from nearby supernovae explosions. The debris left over from their formation still orbits around them as the stars begin to fuse hydrogen into helium inside their cores and then burst into life. This leftover debris of gas and dust is the raw material for planet building. The conditions are not always right for planets to form—not all stars have them. But certainly around the star we call the Sun, a complex system of planets, satellites, asteroids, and comets has arisen.

All of these bodies interact directly with the Sun and each other via several different processes, forces, and events. The entire solar system is bathed in the constant wind of charged particles flowing from the sun, which either interacts with planetary atmospheres and surfaces, or is partially deflected by magnetic fields. Of course, all of the bodies in the solar system have gravity that spans the distances between them. Gravity can allow one planet to hold a satellite in orbit, shred a comet that comes too close to another planet, and create a stunningly beautiful and sophisticated series of icy rings around yet another. And some material from the early days of the solar system is still available, occasionally impacting the surface of one world and thereby throwing more rocks and other material back into space, ready to find its way to yet another.

Planets as Systems

On a single planetary body, all the systems and processes are of course still acting together, influencing one another. On the Earth for example, our biosphere, atmosphere, hydrosphere, and lithosphere all work together over time to cycle carbon. The fact that we have such abundant oxygen in our atmosphere is not because it started out that way, but because of interactions between the atmosphere and biosphere. The Earth's global magnetosphere, created by processes deep within the planet, interacts directly with the atmosphere as well as shields the surface from the high-energy particles streaming from the sun. Life would have developed differently had it not been protected from much of the damaging effects of the solar wind.

Even on what might be seen as a relatively simple world like our Moon, which possesses no hydrosphere, biosphere, or magnetosphere, there are still vast numbers of processes working together. Such processes have been in operation throughout the Moon's formation and evolution. For example, impact events create explosions that throw out rocky debris in the form of crater **ejecta**. This ejecta ages over time. It is slowly churned up by subsequent smaller impacts and constantly altered by charged particles from the sun. The ejecta becomes incorporated into the Moon's soil, the regolith, which owes many of its distinct properties to micrometeorite impacts that would not occur if there was an atmosphere to filter them out. So planets are themselves a group of systems within systems. To understand planets we need to look at the major processes, important events, and key interactions that have affected them and still tie them together.

THE INNER PLANETS

Planets in our solar system can be loosely separated into two kinds, and they each fall into distinct regions. The *inner planets* are the smaller, mostly rocky bodies found closer to the Sun, while the *outer planets* are larger, gas- and volatile-rich planets found further out. The inner planets are also known as the terrestrial planets, or rocky planets. The outer planets are known as the giant planets, with Jupiter and Saturn referred to as the gas giants, and Uranus and Neptune as the ice giants.

The Basics Are Powerful

There are some powerful basic concepts that allow scientists to draw important conclusions about the essential properties of the planets. Some are so obvious, like "Hotter Toward the Sun" that one might wonder how they can be of use to professional scientists. But it is upon simple principles like these that the very first generation of scientific models were made. Some of these principles, such as "Volume versus Surface Area" are dealt with in specific chapters of this book. A few are discussed here to introduce the inner planets and the pivotal ideas that allow us to understand them.

Hotter Toward the Sun

This basic observation has resulted in complex models about the formation of the planets and the solar system as a whole. For example, why are the rocky planets found in the inner solar system, and the gas and ice-rich planets found in the outer solar system? The condensation sequence is a model of planetary formation that suggests that only heavier refractory materials like rock and metal could condense in quantity from the solar nebula near the Sun, while in the outer solar system, lighter, more **volatile** elements such as water and methane could condense. Jupiter grew quickly enough that it was able to gravitationally hold large amounts of hydrogen before the bulk of light elements was swept from the solar system by the **solar wind**. The concept of "Hotter Toward the Sun" is also in play as scientists try to understand more detailed differences between the planets. Why is Mercury so rich in metals compared to the other inner planets? One theory suggests that its close proximity to the sun meant that more of the lighter weight materials were vaporized. Why is Venus, our supposed sister planet, so stiflingly hot? Again, one theory suggests that it was just enough closer to the sun than we are that the atmosphere experienced a runaway greenhouse effect.

Of course, any models that develop around the "Hotter Toward the Sun" principle can be misleading if planets and other bodies are no longer in the locations where they originally formed. The early solar system was a very dynamic place, and bodies were running into each other and having gravitational interactions that altered their orbits much more than we see today. Additionally, models created using this principle can be incorrect if the body in question was altered by some other process, such as a giant impact event. Scientists must always be aware that the conditions they are currently observing, and around which they build their theories, may have been different in the past. Looking for clues about how conditions have changed with time is a critical aspect of planetary science.

Superposition—What's On Top Is Youngest

This amazingly simple idea is the basis of modern stratigraphy and is responsible for our ability to estimate the ages of different geologic and biologic events on Earth (or any planet). Rocks on the Earth are in layers because the processes that create rocks, such as sediment emplacement, extrusion of lava, dust and ash fall, the build up of limestone or coral reefs, and more, all happen on top of one another. This builds up a stratigraphic sequence of rocks over time. Imagine first there are several lava flows, and between them are layers of ash that settled out of the atmosphere between eruptions. Then the land was covered by a shallow sea, and over time the bottom of this sea was covered with shells and such that compacted to form a layer of limestone rock. The sea retreated and a large amount of airborne sediment moved through, creating dunes. The sand from these dunes was eventually cemented together by pressure and other processes to form sandstone rock. Then into this sandstone an impact crater formed. Last, a river eroded down through all of these rocks and exposed them in a deep cliff face. A scientist studying the stratigraphic sequence in this cliff face does not have to have been present to know which events happened first, then in the middle, and then last. The top is where to look for the youngest events.

Again, this powerful idea has limitations. Rock layers can be overturned, even bent, mixing up the sequence. On a planet like Mars, wind can deposit several layers of dust and ash that protect the rock beneath from impact-cratering events. And then later the wind regime can change and erode the layer away completely. Scientists faced with this situation may underestimate the age of the rock layer underneath, not knowing it was protected by another layer. All history of the intermediate dust layer might even be completely lost.

Bulk Density—Density Estimates Composition

The bulk (average) density of a body is an extremely basic piece of data. And yet it is critically important to making estimations about the composition of the body and the nature of its interior. Planets are formed mostly of volatiles like ice (and little of that in the inner planets), rocky material, and metals. The density of water/ice is 1 g/cc (gram per cubic centimeter); the density of a basic silicate rock is 2.7 g/cc; and of iron metal is 7.9 g/cc. Assume we find a planet with a rocky surface with no metals and a bulk density of 5 g/cc. We know immediately that the planet is not composed of rock the whole way through. The density is an indicator that there is metal inside this world, and even gives us an approximation of how much metal.

Again, there is a twist to the technique. Some worlds are so large that they are compressed under their own gravity. Earth, for example, has a bulk compressed density of 5.5 g/cc. This is what a spacecraft in orbit of the Earth would measure. Normally, rock and metal the size of the Earth wouldn't actually be that dense. The planet's gravity has squeezed the materials into a smaller space. To compensate for the gravity of larger worlds, the effect of gravity can be removed. This uncompressed density is the proper density to use when looking to estimate the composition of a planetary body.

TRICKS OF THE TRADE

As noted in the previous section, even simple principles can have limitations and complexities to be considered if they are to be used properly. Planetary scientists use a host of instruments and types of data to build up their ideas and hypotheses. Each of these is subject to its own limitations, and knowing when to use what instrument, or where certain data are the most reliable, is part of what it means to be a planetary scientist.

The most widely used technique of the scientist is not related to a specific instrument, it is the development of scientific models. Scientific models are the way professionals represent or explain a system, phenomenon, or process. When most people hear "models" they might think of model trains or modeling clay. A model train is a small version of a regular train. It isn't accurate in terms of size, speed, or composition, but it does allow you to see that trains have cars, that they move on tracks, and that they can carry people or cargo. A model train is a form of scientific model about trains, accurate in some respects but not others. If a scientist is interested in showing how trains move, a small model train might be what is needed. Modelling clay is so named since clay can be used to sculpt forms. You can mold a small Earth out of clay and draw upon it. It will not be correct in size, composition, or color, and it won't have an atmosphere, but it will show that the Earth has surface features and their relationship to one another. Again, it does not show all aspects of the Earth system—no model can do that. If a scientist wants to show how the continents are spaced apart, this model may suffice. If a scientist wants to show the circulation of atmosphere near a hurricane, another model, perhaps a computer simulation, would be necessary.

Models are constructed based on the best available data. It may take many models to describe one system, or models may be in competition with one another, with scientists debating which more accurately depicts the part of the system that is in question. As scientists make more observations and collect more evidence, they change their models to be more accurate. Sometimes new data comes to light that makes an old model completely obsolete, such as the old Earth-centered models of the solar system. Then the old model is essentially thrown out completely, and a new one is adopted as the current "most likely" model.

It is important to note that scientists do not "believe" or "disbelieve" a certain theory, model, or fact. Scientists postulate, consider, think, hypothesize, estimate, and approximate. They do not "believe" even if they use that term (which they often do, in error). A scientist may prefer one model over another because it is more complete, more robust, and better explains the available data. If tomorrow different data come to light, then the same scientist may choose another model as the most likely explanation for the phenomena. The general public often views this as the scientist being "wrong" at first, and it may lead them to think the entire process is shaky. This isn't the case, since there isn't a "wrong" or "right" model. Instead, there are those that fit the data, have met more tests, and have been verified more often, and there are those that have not. Even the very best model can become obsolete overnight. This is not a reason to reject the scientific process. Rather, it is the evidence that the scientific process is working, and that people are not holding on to old ideas just because they "like" them.

A great deal of the data that are generated in planetary science, and therefore fed into the models scientists are creating, are image data. Spacecraft send back thousands of images of planetary surfaces, atmospheres, magnetic fields, and more. It can be very confusing, even for professionals in another field, to immediately interpret an image of any type of astronomical phenomena. When looking at images, even those in this book, first consider the wavelength that the image was taken in, and the scale of the picture. You may be taken off guard if you think you are seeing a rough planetary landscape, and instead are looking at a microscopic photo of the surface of a dust grain. When not noted, most images are either simple black and white or color images in regular "visible" light. But wavelength is not always specifically noted. Check captions and other details to see if instead you are seeing a radar image, or one in infrared, or even ultraviolet light. Planetary sources can be imaged in all wavelengths of light, and each wavelength tells us something different and important about the target. Interpreting them properly is important. Radar images of the surface of Venus, for example, have dark and light areas. But these do not mean that if you were there in person you would see darker and lighter areas. When imaging in radar, the bright areas are rough on the scale of the radar used, and dark areas are smooth.

Another important point is that with the advent of easy-to-access and easy-use image processors, it is simple enough to create an artist's impression or interpretation of an object. These can be very difficult to distinguish from real data. Always check captions, and be skeptical if something looks amiss. The classic picture of a galaxy with a "You Are Here" arrow is famously misleading. No spacecraft have been beyond the next star from the Sun, let alone gotten all the way out of the galaxy to take such an image. These are pictures of other galaxies thought to resemble ours, and have an arrow showing where we would be in relation to the center, IF that were in fact our galaxy.

QUESTIONS TO BE ANSWERED

Each chapter in this book mentions some questions and issues still under investigation. There are more interesting problems awaiting investigation in planetary science than there are people available to investigate them.

One of the most pivotal is the relationship between sample data (for example, a rock you are holding in your hand) and remote sensing data (for example, images taken from telescopes or spacecraft). It is obvious that sending spacecraft to the planets is expensive. Much more expensive is sending people there to collect samples with complete understanding of their geologic context, and then bringing both the samples and the people back. It is much less expensive to use robotic probes to take pictures or collect other data from a planet remotely, and then just beam the data back to Earth. But remote sensing data can be very difficult to interpret. Are the rocks the composition we think they are? Is dust cover obscuring what is underneath? Has influx from the solar wind changed the upper rocks, but not the lower ones? In essence, how well do our techniques for modeling our remote sensing data really approximate what we would learn if we had a sample of the rock right in our hands?

Because of the processes mentioned, among others (dust cover, water erosion, space weathering, micrometeorite impacts, solar wind influx) planetary surfaces are constantly changing with time. Some of these processes will change a pristine rock in one way, and some of them another. When we see the rock from space, scientists must attempt to figure out if they are actually seeing an unaltered rock, or one that has indeed been altered, and then unravel which of the processes might have affected it. This complex endeavor is currently at the crux of many investigations in planetary science. For best use of resources, scientists need to use remote sensing as much as possible. But having rocks returned from the Moon during the *Apollo* program has shown us just how useful collecting rocks with geologic context can be. Scientists need to learn much more about how processes change planetary surfaces to make the remote sensing data as useful as possible. To do this, some sample data will have to be obtained.

Planning how much and from where rocks from other planetary bodies will be gathered is a major effort of the space program.

FOR MORE INFORMATION

The Planetary Society provides a popular-level introduction to the inner planets at http://www.planetary.org/explore/topics/compare_the_planets/terrestrial.html.

NASA stores a large collection of mission image releases, both for the inner and outer planets, at the Photojournal (http://photojournal.jpl.nasa.gov/).

While all of the terrestrial planets in our solar system have been discovered, astronomers are beginning to consider finding them around other stars. The Terrestrial Planet Finder (http://planetquest.jpl.nasa.gov/TPF/tpf_index.cfm) is a NASA mission concept under study to do just that.

2

Cosmic Collisions: Cratering

INTRODUCTION

The Solar System is a busy place, and in spite of the vast distances between objects, they do run into each other from time to time. Have you seen shooting "stars" in the dark nighttime sky? These are not stars falling to Earth. Stars are thousands of times bigger than a planet, so this old idea does not make sense. In fact, what you are seeing are particles no bigger than pebbles, or even dust, burning up as they collide with our upper atmosphere. The vast majority of these objects never make it to the surface, but instead vaporize in a streak of light that we all know well, and perhaps have wished upon.

Tiny dust-sized particles litter the inner solar system, but as you look for larger and larger pebbles, and then rocks, there are fewer and fewer of them. Still, there a billion objects the size of a car, and more than a hundred thousand the size of the Statue of Liberty inside the orbit of Mars. There are thousands more than a mile across. If you want to find tens of thousands of larger rocky bodies, look for them in the asteroid belt. This is a wide expanse of small, rocky objects orbiting the outside of the inner solar system, between Mars and Jupiter. Of course, the largest rocky objects in the inner solar system are the planets themselves.

When any two objects run into one another, that is referred to as an **impact event**. These can be quite small and unremarkable; a dust grain hitting the surface of the Moon, causing a tiny bead of melt to form; or as mentioned, a pebble striking the atmosphere of Earth, and disappearing in a flash of light. However, impact events can also be dramatic, even catastrophic. If two objects of a similar size collide (or if one is a little smaller and moving very, very fast) the objects might completely destroy each other in a massive explosion.

But given the numbers and sizes of objects and the vastness of space, collisions between two large objects are rare. Instead, it is more common for a large object like a planet to collide with any number of smaller objects in a range of sizes. Scientists generally refer to the larger object as the **target** and the smaller object that collides with it as the **projectile** or **impactor**. When two objects of very different sizes collide, the impactor is usually destroyed completely. The target usually survives, but with a pockmark or scar to bear witness to the impact event. This circular mark on the surface of the target is called an **impact crater**. Impact craters can range in size from a centimeter or smaller to hundreds of kilometers across, depending on the size and speed of the projectile that formed it, as well as the composition of the projectile and the target.

Impact craters are some of the most important features on planetary surfaces. They are found on every inner planet, and on every asteroid and moon we have yet imaged. They provide the most accessible and reliable means for looking back into the deep history of the solar system to a time when impact events were much more common. Impact processes have shaped the surfaces of all the inner planets, allowed for the formation of the Moon itself, and may have played a key role in the existence and possible extinction of species of life on Earth.

CRATER FORMATION

The basic idea for the formation of a crater is straightforward: an impactor hits a target, and the resulting explosion forms a circular-shaped depression with a raised rim and surrounding blanket of ejected material. The formation of the crater is driven by the energy of the impact, largely the kinetic energy of the impactor, which is usually moving very fast relative to the target. There is a vast amount of energy in an impact event, so much so that the formation of a crater is best modeled as an explosion. Craters are not formed by a projectile "digging out" or pushing aside rock and dirt to form a hole in the ground. In fact, the kinetic energy from an impact creates an explosion almost exactly like a bomb going off.

Impact crater formation can be envisioned as a three-stage process. Stage one is *contact and compression*, stage two is *excavation*, and stage three is *modification and collapse*.

1. **Contact and Compression**. During this stage the impactor makes contact with the target. The energy from this projectile couples (transfers) to the target, and begins to compress and decelerate the impactor at the same time it compresses and accelerates the rocks of the target. This compression happens at speeds that exceed the speed of sound in rock (**supersonic**)

creating a shock wave through the target and projectile material. As the compression continues, target rocks suffer melting and perhaps vaporization. The entire contact and compression stage can take much less than a second.

2. **Excavation**. In the second stage the transient or initial crater-shaped depression begins to form. This initial crater will undergo substantial changes before reaching its final state. The transient crater rapidly expands in a hemispherical shape as material is compressed downward. Part of the energy of the expanding shock wave rebounds in a **rarefaction wave** that forces material to be thrown outward. The ejected material blows outward in a cone-shaped curtain that falls to the ground to form a blanket around the new crater. As this occurs, material is actually flipped upside down to form a raised rim around the edge of the depression. In more energetic events, cracks and fractures form under the crater, down into the target material. This stage can last from a few minutes to a half an hour or more.
3. **Modification**. In this last stage, it is common for some slumping of the crater walls to occur, with material settling into the bottom of the crater, and terraces forming into the sides of the crater walls. A central peak or peak ring may form as rocks at the bottom of the crater resist compression and deformation. These rocks may rebound back upward at the point of impact to form the central structure of the crater. The last of the ejecta is emplaced around the crater, as well.

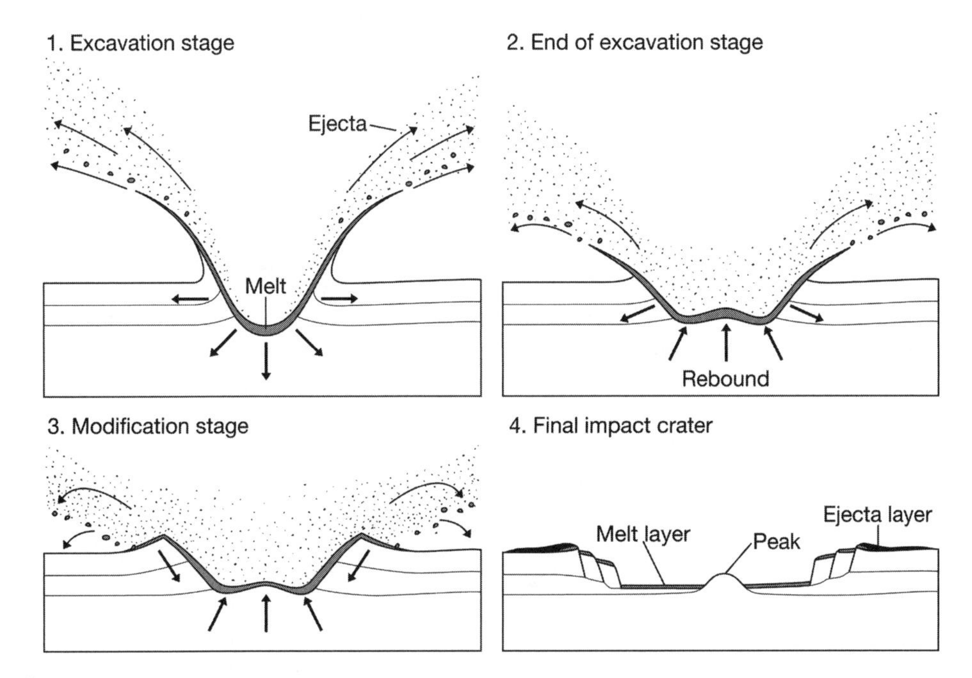

Figure 2.1 Crater formation has several stages, from excavation to final structure. The amount of time required for crater formation depends on several factors, including size of impact and the surface gravity of the target. The formation of a complex crater is shown above, simple craters have the same stages but have less rebound and consequently do not have central peaks.

Impact Craters and Volcanic Craters

In planetary science, the word "crater" almost always refers to impact craters. There are other types of craters, and for earth scientists impact craters are relatively rare compared to volcanic craters. The fact that volcanoes also leave craters led in part to a famous disagreement between two American scientists of the 1960s. Gerard Kuiper pointed to the lunar craters and interpreted them as evidence of volcanoes, arguing that the Moon was once molten. Harold Urey argued that the craters were most likely due to impacts, seeing little evidence of craters like terrestrial volcanic craters and suggesting that the Moon formed cold and remained relatively cold. The *Apollo* missions showed both men were right, and that both were wrong. The vast majority of craters on the Moon are due to impacts, but geochemical studies of the returned lunar samples showed that the Moon was once molten.

At this point, the crater has essentially found its "final" shape. But modification does not ever really stop. Craters change with time the same way any geologic formation changes. Craters on the Earth are eroded by rain and wind, are covered by vegetation and human activity, are deformed by earthquakes and landslides, are filled in by lava flows, and eventually are utterly obliterated. Even on a relatively quiet body like the Moon, craters change over time, suffering from landslides, impacts of other craters on their rims, and in the past, infilling by lava.

In fact, the processes that slowly modify craters over time are well-characterized for the Moon and can help indicate the relative ages of craters, as long as they are of similar size. Smaller craters will otherwise "age" much more quickly than larger ones.

Figure 2.2a

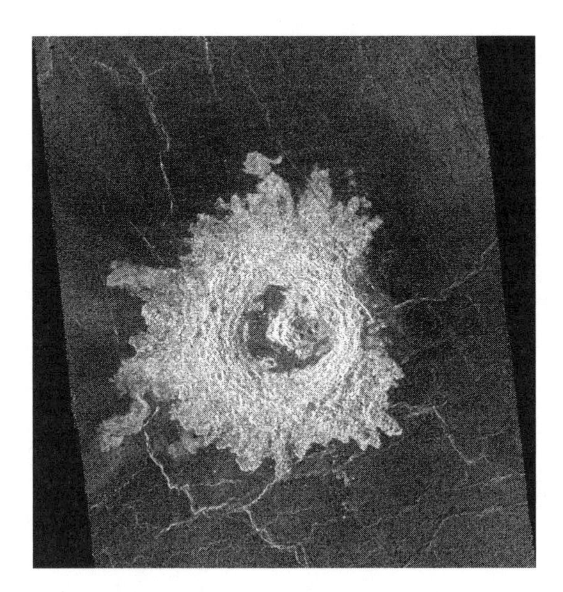

Figure 2.2b

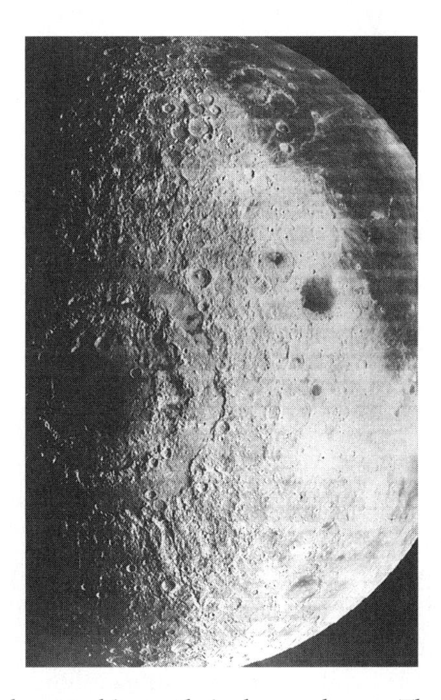

Figure 2.2c As craters become bigger, their shapes change. The smallest impacts have bowl shapes, and relatively smooth interiors. These are called simple craters, and include Barringer Crater, seen in Figure 2.2a (previous page). Larger impacts create complex craters, which have structures in their interiors, often including a central peak or ring, like the venusian crater in panel b. The largest impacts create basins hundreds of kilometers across. These can have very complicated shapes and several different rings, as we see in panel c, showing the Orientale Basin on the Moon.

CRATER COMPARISONS

Classifying Craters

Craters are not identical. The most critical differences are a result of the energy of the impact event that created them. Lower-energy impacts create small, simple craters, while higher-energy impacts create larger craters with significant modification. The highest-energy impacts are those that formed the large basin-sized craters on the inner planets. Such craters are so large that the curvature of the target planet plays a significant role in the modification of the crater as it reaches its final form.

Simple craters are shaped like small bowls. They have an even, raised rim, smooth interiors, and sharp features. In all cases, simple craters are the smallest craters to be found on a target planet. **Complex craters** are the medium to large-sized craters found on a target planet, and as their name suggests, their shape will be less easily described than simple craters. They can have ridges or terraces on the inside of their rims. The floors of such craters will be flatter, and may have a central peak or a peak ring structure. The rim and interior will have more slumping, with possible debris slides along the walls.

The largest craters to be found on the inner planets are called **impact basins**. The shapes and forms of such craters are very complicated, and they may not look like typical craters in any way. Instead of a neatly defined rim, impact basins have multiple rings of rims, with flat floors between each ring. Impact basins can be hundreds of kilometers in diameter.

Primary versus Secondary

An impact can cause more impacts, and in fact, they usually do. The more energetic the impact event, the more **secondary craters** will be formed. A **primary crater** is the initial event, usually envisioned as formed by a single impactor striking a target and creating an explosion. But this explosion in turn launches more impactors outward from the primary impact site. Some of these can be small and very low-energy, forming chains of secondary craters near the larger impact site. Because they form with lower energy, they do not usually resemble primary craters—they often have less circular rims, are more shallow, and may form in strings or clumps. But some ejected material can be in large, coherent blocks and thrown a great distance from the crater with very high speeds. The secondary craters formed by this ejected material can be very hard to distinguish from other, smaller primary craters.

Crater Ejecta and Rayed Craters

What happens to the material that used to fill in what is now a big hole in the ground? Some of it is vaporized, some is thrown outward, and some is

Primaries and Secondaries

The combination of lunar samples and crater counting has provided a good understanding of the ages of parts of the Moon. In principle, if one can connect a cratering rate with an age for the Moon, that could be used to determine the ages of surfaces on other objects as well. This would require knowledge of the different production functions throughout the solar system, and how different situations might affect the crater population (for instance the atmosphere of Mars versus the atmosphere-free Moon), but it is theoretically possible. Scientists have produced calculations for Mars predicting crater populations for regions of different ages. Over the last 10 years, the explosion of imagery of the Martian surface provided by the *Mars Global Surveyor*, *Mars Express*, and *Mars Reconnaissance Orbiter* missions has allowed ages to be calculated for very small areas using very small craters.

However, these calculations are controversial. Secondary craters need to be removed from crater counts because they do not follow the same patterns as the production function. Furthermore, they do not accumulate at a regular rate. At large sizes, geologists can make convincing arguments that a particular crater is a primary or a secondary crater, lending some confidence to the crater counts. At small sizes, however, recent research has shown that it can be difficult or nearly impossible to distinguish between primary and secondary craters. A very young area that unluckily got hit by secondary craters from somewhere else could look much, much older than it actually is. Research is ongoing to try to assess the conditions under which "contamination" from secondary craters is a problem.

compressed downward. The material that is thrown out of the crater is called **ejecta**. As one might expect, the ejecta closest to the crater was thrown out with the lowest velocity, and the ejecta found furthest away was thrown out with the highest velocity.

The ejecta closest to the impact crater are found as part of the crater's raised rim. About half the rim height is due to uplift from the impact event, where pressure from the center of the growing crater breaks and forces the rock upward. The other half of the height is made from material blown out from the crater. This close to the impact, the ejected material is thrown out with low enough speeds and energy that it keeps its shape, and if there were layers in the rock, they may survive more or less intact—except they will be upside down. This is because the explosion blows the material out as a curtain or flap, which overturns and then overlays the surrounding terrain. Geologists refer to the overturned layers as **inverted stratigraphy**, stratigraphy being a geological term referring to layers of rock.

Ejected material is thickest near the rim of the crater, and thins with distance. Within about one crater radius the ejecta forms a thick and relatively even deposit of hills and "hummocks." This is known as the **continuous ejecta blanket**. Past the distance of about one crater radius is the

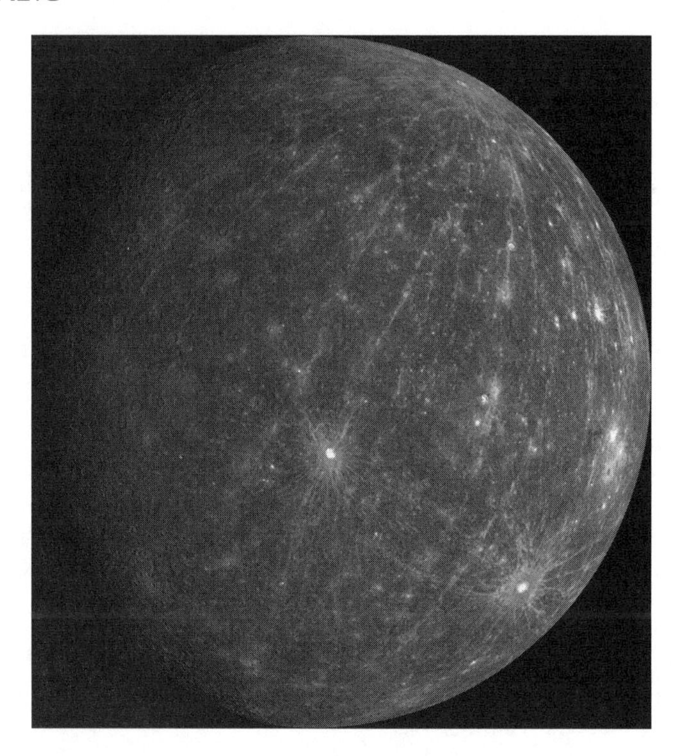

Figure 2.3 This image of Mercury taken by the *MESSENGER* spacecraft shows several rayed craters. In some cases, rays can stretch for extremely long distances, as with the crater near the upper left of the image whose rays can be followed nearly the entire length of the planet.

discontinuous ejecta, where the ejecta is broken up into radial lines and arcs of debris and secondary craters.

Newly formed impact craters are often surrounded by bright **rays**. Rays extend the furthest away from an impact site of any ejected material (not counting the material that has been thrown right off of the planet, of course!). How they are formed is not well understood, but certainly a small amount of material moving at high speeds must be involved. Rays are not actually blankets of ejected material, but are caused by high-energy material altering the surface rocks already on the planet. In the case of the Moon, crater rays are bright (high albedo) lines that can extend halfway around its diameter. The underlying "soil" of the Moon contains small bits of melted glassy material, which darken the surface. Energetic material blasted from an impact event somehow shatters and mixes the upper soil of the Moon, making it bright along neat lines away from new impact craters.

Once a crater is formed, the rays are the first feature that will fade with time. Bigger impact events create larger and more long-lived rays, but rays are still a relatively short-lived feature of any crater. Ejecta blankets go next, with thinner ejecta being erased first, and continuous ejecta and secondary craters last. Then the raised rim of the crater will be worn away. Eventually the crater may be obliterated completely.

Depth to Diameter Ratio

Simple craters have a depth to diameter ratio of approximately one to five. The depth of a crater is measured from rim to floor, and diameter measured from rim to rim. This is consistent from planet to planet in the inner solar system, and exceptions to this ratio are very useful in spotting important differences between the planets or the "target" materials. Even small craters on Mars can be unusually "shallow," which leads many scientists to believe that the target material was icy or even wet at the time of crater formation. Craters can also be filled in by sand and lava, which changes their depth to diameter ratio.

Simple to Complex Crater Transition

Comparing craters from one planet to another gives scientists some important insight into the factors important to crater formation. Simple craters differ from planet to planet. The smallest craters on a planet will always be simple ones, but when you look for craters of increasingly larger diameter, there will be a point at which you no longer find simple ones, only complex craters. The diameter range where you begin to see complex craters is called the **simple to complex transition** and this diameter is different from world to world.

A first look at Figure 2.4 suggests that planet size is the key factor in determining the simple to complex crater transition diameter. A second look shows, however, that the transition diameter range for Mars and Mercury overlap. Mars is much bigger than Mercury, so how can this be?

Planets are not all composed of the same amounts of materials. While the inner worlds are predominantly rocks and metals, with some lighter materials like water, they do not have exactly the same fraction of each. The Moon has no (or a very tiny) metal core. Mercury is a planet with a very high average density (mass per unit volume) because it has a large metallic core. The result is that the surface gravity of Mercury is similar to Mars, a larger planet but one with a lower average density. If we look at the transition diameter as a function of gravity for the inner planets, we can see a much stronger correlation.

This correlation between gravity and transition diameter is telling us something important about the factors that affect crater formation. Gravity is a very important variable. As a crater forms, the surface gravity of a planet plays a key part in how the crater develops and its subsequent modification. On low-gravity planets, it is easier for a crater to develop

Object	Transition Range (km)	Object Radius (km)	Surface Gravity (m/s^2)
Moon	9.2-13.3	1737	1.6
Mercury	4.2-5.4	2440	3.7
Mars	3.2-4.4	3390	3.7
Venus	1.7-3.7	6052	8.9
Earth	1.5-2.5	6371	9.8

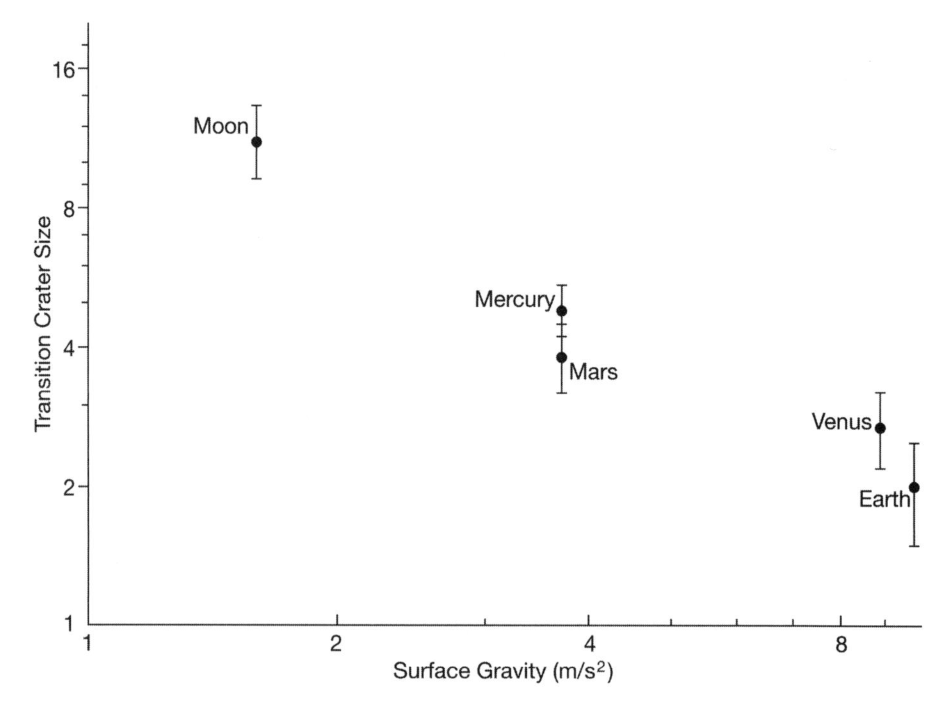

Figure 2.4 Each body has a critical size (called the transition diameter), below which only simple craters are found and above which complex craters are seen. Using the data in the table (top panel), we see the transition diameter (measured in km) decreases as an object size increases. However, the surface gravity on a body is an even better predictor of transition diameter than the objects' own diameters.

and maintain a smooth, simple bowl shape, even at larger diameters. The higher the surface gravity, the more stress (force) is exerted on the forming crater, and the more likely the rocks will strain (break) and undergo modification, even at lower crater diameters. This is a somewhat simplistic view, but it provides insight into the critical variables related to crater formation.

This plot shows us something else—Mars does not appear quite where one might expect. If surface gravity were the only variable, then the transition diameter for Mars should be exactly the same as for Mercury. But Mars generally has a smaller transition diameter than Mercury. This is because gravity is only one of many variables. The nature of the target rocks is also important. A look at the depth to diameter ratios for some Martian craters also shows anomalies. This fact and other lines of evidence suggest to scientists that some of the target rocks on Mars contained liquid water at the time of the impact event. Some craters on Mars look more like mud splashes than scars from impact events.

CRATERED SURFACES

Craters are everywhere in the inner solar system. They are one of the few features common to all the inner planets, as well as to the asteroids and satellites. Because they are on all the planets, and because there are large numbers of them available, scientists can learn a great deal about the history of a planetary surface by looking at the impact craters. In particular, they get great insight by studying the variation in size, depth, density, degradation, spacing, number, and shape of the craters.

Crater Density

Cratered surfaces range from those that have no visible craters at all, to those that are so densely cratered that no new crater can form without wiping out craters that came before it. Impactors run into the planets from all directions, randomly striking their surfaces. So all the inner planets should be covered with craters—unless there are processes at work that destroy them, either now or in the past.

The Moon gives us important clues about how impact craters have formed over time throughout the inner solar system. The Moon is covered with craters of all sizes. The most densely covered areas are completely **saturated** with craters. If a new crater formed, it would be hard to tell since there is no empty space left in such areas. Yet when we look at the Earth, we do not see anyplace with a crater density that high. What is responsible for the difference?

We know that fewer craters are being formed now than in the past, but there are still impact events in the inner solar system. Since there is no reason that projectiles should prefer to strike the Moon, rather than the Earth, it is highly likely that something has wiped out the terrestrial craters.

Knowing what we do about the Earth, this isn't difficult to believe. Wind, rain, human activity, vegetation, earthquakes, volcanoes and much more all serve to change the Earth's surface. Some of these processes work relatively quickly, like a massive flood that redirects a river or fills a valley with silt in a matter of days. Other processes are relatively slow, like the subduction of continents and the extrusion of new oceanic crust on the ocean floor over the course of millions of years. But from a geologic perspective, both of these processes are very rapid compared to the billions of years that the Earth has been around. Comparing the crater density from the Moon to the Earth was one of the key ways scientists were able to confirm the extent and activity of Earth's many dynamic processes over time, particularly plate tectonics.

Craters are the indicator—if you see a surface that is not completely covered with craters, then something has happened to that surface. It may have happened long ago, when the planet was young, or it might have happened relatively recently in geologic terms, but something altered the surface to remove the craters. It might be as simple as craters formed in a mountainous region being obliterated by landslides and other slow downslope movement, or it might be far more complex. There are regions on Mars that have no craters; these might have been covered with dunes, filled in by lava, or washed out by ancient water flow. Or perhaps all three processes occurred.

Crater Counting

Planetary surfaces are giant collecting areas for craters. As time goes on, more craters form until something like a lava flow comes through and wipes them out. Afterward the lava hardens, creates a new surface, and the craters start building up again. If one knew how fast craters formed, then it might be possible to find out when the lava flow occurred by counting the craters that have formed on it since it hardened.

In fact, conducting statistical studies of the size and number of craters on a given surface is a straightforward way to find relative ages. Barring other factors, a surface that has a higher crater density is older than one with a lower crater density. That is, the older a surface is, the longer it has had to collect craters. Younger surfaces have been collecting craters for a shorter amount of time. This study is referred to by the fitting, if unglamorous term "crater counting."

The lunar surface was initially covered with craters. After the craters formed, the low-lying areas of the Moon were filled in with wide, thick expanses of lava that obliterated the impact craters. These areas are the **mare**, and after the lava cooled, they began to collect craters once again. Counting the number of craters at a given diameter gives scientists a way to quantify the relative difference in ages between different surfaces. Even if the relative difference cannot be seen easily by just looking, as with these images, crater counting can provide the means to distinguish older and younger surfaces from one another.

For the Moon we have an important, added bonus. We have brought back rocks from the surface of the Moon and have been able to determine their ages independently using radiometric dating techniques (see the chapter on Rocks as Clocks). For those rocks, we know their ages absolutely. Scientists have been able to tie these absolute ages to some surfaces whose crater densities have been determined by crater counting. This has allowed

The Terminal Lunar Cataclysm

The samples returned from the Moon by the *Apollo* astronauts and two Soviet robotic missions contained a number of surprises. One of the notable ones involved the ages of the largest basins on the Moon. While the Moon itself formed 4.5 billion years ago, there was little evidence for any basin-forming impacts older than 3.9 billion years, with a large number of impacts indicated at roughly 3.9–3.8 billion years ago. Given the conception of a smoothly dropping impact frequency from the end of the late heavy bombardment, scientists didn't expect any cluster of impacts, and certainly would not have predicted any cluster to fall as late as the one seen.

A suggested solution to this discrepancy is termed the **Terminal Lunar Cataclysm** (or TLC). Rather than the smoothly falling impact rate, this scenario includes a relatively brief period of increased impacts before the rate fell back to the pre-"cataclysm" rate. Skeptics of the TLC model argued that it was difficult to generate a large population of impactors 600 million years after the planets formed, and that the apparent clustering of ages was more likely a result of poorly understood data.

Within the last 10 years, however, further evidence has mounted in favor of the cataclysm idea. Meteorites from the far side of the Moon are also consistent with the TLC, and dynamicists studying the behavior of Jupiter and Saturn early in solar system history showed that their mutual gravity could have generated a large shower of impactors at roughly 3.8–3.9 billion years. The idea is not yet settled, and remains of great interest to lunar scientists.

for the lunar crater statistics to give an absolute age difference, not just a relative one.

The crater counting data for the highlands plots higher than for the mare, showing that the highlands have a higher number of craters at each crater diameter. In addition, dates determined by radiometric dating methods confirm the results of the crater counting. Because we have these absolute dates, we can estimate the absolute age of other cratered surfaces on the Moon, and construct a more accurate history for that world.

The lunar cratering record has been extrapolated to other planets. It is difficult to do, since we know an impact event of a given energy will form a different crater on each world, due to variations in target material, surface gravity, etc. However, estimating these differences allows for scientists to put some constrains on the age of surfaces throughout the inner solar system.

Crater Production Over Time

Looking at cratered surfaces throughout the inner solar system, but predominantly on the Moon, scientists have been able to construct an idea of how impact events have proceeded during the lifetime of the planets.

In the early solar system 4.5 billion years ago, there was a great deal of material, in the form of asteroids and comets of various sizes, that had not yet been collected by the planets, or had not yet been thrown out of the solar system or into the Sun by the gravity of one of the giant planets. This material was free to collide with itself or with the newly formed planets. The era of the major basin-forming events on the Moon might have occurred during the end of this period of **late heavy bombardment**, or the basins might represent a spike in the cratering rate—a sudden influx of large projectiles into the inner solar system. Either way, the production of craters dropped off steeply after about 3.8 billion years, and appears to have continued to decline to the present day. This is to be expected as material gets swept up by the planets, smashed in collisions with other material, or thrown out of the solar system completely.

However, a declining cratering rate does not mean that no impact events have happened in recent geologic history. Barringer Crater in Arizona formed recently in geologic time, only 40,000 years ago. This crater is about a mile across, and would have been highly destructive in its time, blasting down forests, killing wildlife, starting fires, and raining down rock over a huge radius. And a mile-wide crater is considered small by planetary terms.

One of the reasons scientists study impact craters is to create better models of the change in the impact cratering rate over time—has it fallen off smoothly, or are there still spikes continuing today that might imply more impact events for Earth in the future? A sizeable impact on the Earth today would be quite dramatic at best, and tragic at worst. The Earth has seen massive impacts in the past that contributed to the extinction of whole species.

CRATERS AND EXTINCTIONS

Scientists have compiled a great deal of evidence suggesting impact events are responsible for climate change on the Earth, with effects varying from temporary to very long lived. Certainly a large impact event would have far-reaching effects; lifting dust into the upper atmosphere, releasing greenhouse gasses, creating massive shock waves, obliterating vegetation by blast or fire, as well as possibly vaporizing large quantities of water and creating tidal waves. Such events could have been responsible for extinctions as seen in the fossil record of Earth's rocks.

The most famous extinction event is the one that caused the extinction of the dinosaurs, as seen at the end of the Cretaceous era, called the "K–T boundary extinction." Many theories were put forth as to how the dinosaurs, along with about 85 percent of the species on the planet, could have died at the same time about 65 million years ago. In the 1980s, an impact hypothesis for their extinction was suggested. While a massive impact might be quite enough to deal with on its own, the dinosaurs may have already been struggling to adapt to a changing environment. In this case, a huge impact event would have been more than enough to push them over into extinction.

This hypothesis was not considered a likely explanation until scientists identified the crater that was formed in this event. The crater is called Chicxulub, and is located near the coast of the Yucatan Peninsula of Mexico. The crater is a multiring basin at least 180 km across, and possibly larger. The exact size of the crater is difficult to determine since it has suffered so much modification that it is hard to identify the rims at all. Half of the crater lies under water in the Gulf of Mexico.

This is by no means the only extinction event that has occurred on Earth. Several events, including the largest, the Permian–Triassic extinction, have been recorded by marine fossils. In the P–T event, more than 95 percent of marine species were wiped out, and it took more than 100 million years to bounce back to previous levels of biodiversity. It is possible that impact events have played a role in these extinctions. Scientists continue to study the timing of the extinction events. Do they happen in a predictable, periodic manner? Are they random? Do they seem to coincide with spikes in the impact cratering rate? Not enough is known to definitively answer these questions. More data of higher precision is required to tie down the dates of impact and extinction events more specifically. But certainly the idea of extinctions on Earth possibly being tied to impact events is a highly motivating reason to learn more about crater production, here and throughout the solar system, and astronomers have increased the capabilities of their impactor search programs throughout the last decades as a result.

THE GIANT IMPACT

As noted, impact events come in all shapes and sizes. If the event is extremely energetic, then no crater may result because the impactor and the target both are destroyed or melted to the point that no crater remained.

For many years, some specific characteristics of the Moon puzzled planetary scientists. There did not seem to be one coherent model that explained the data. A change in the way scientists viewed the last stages of planetary accretion in the mid-1970s allowed for a new model to emerge. This model is the **Giant Impact Hypothesis** for the origin of the Moon. In that model, the last stages of planetary accretion are dominated by very large almost-planets called **protoplanets** colliding with one another. There may have been a dozen such protoplanets in the very early inner solar system, with a final stage of collisions bringing us to our current total of four, and one huge Moon.

The Giant Impact Hypothesis suggests that the proto-Earth was struck by another protoplanet possibly as large as present-day Mars. The resulting impact event would have caused massive melting and vaporization of significant percentages of both protoplanets. Vapor from the event found its way into orbit of the Earth where it cooled and condensed into dust, and accreted to form a large satellite in Earth's orbit. The Giant Impact Hypothesis underscores the importance of understanding impacts at all sizes, and how cosmic collisions have played a critical role in the development of the planets, as well as life on Earth.

CONCLUSION

Impact craters can be found throughout the inner solar system, and they are the storytellers of an amazing past. The distant past, 4.5 billion years ago, is filled with cosmic collisions that started out as massive bodies the size of the Earth and Mars smashed into each other. And while impact events in the recent past are vastly less energetic, they are by no means unimportant to us today. Extinction events such as the K–T boundary extinction may have been caused by a massive impact event. And as recently as 40,000 years ago a crater a mile across was blasted into the Arizona desert.

Scientists study these events by looking at the impact craters they leave behind. These impact craters are of all sizes, and pepper the surfaces of the planets. Cratered surfaces can be examined using crater counting techniques to understand their relative ages. Rocks from impact sites can be studied chemically to determine radiometric absolute ages. Individual craters can be examined to see how they have weathered or been modified over time. Comparisons between craters and the cratered surfaces from one planet to the next allow us to look back into the past, and put together part of the complex story of the formation of the solar system and the evolution of life on Earth.

FOR MORE INFORMATION

The classic technical consideration of impacts and impact processes is found in *Impact Cratering: A Geologic Process* by H. J. Melosh (Oxford University Press, New York 1989, 245 p). Although out of print, those interested in details of the physics of impacts should find a library copy.

The "Explorer's Guide to Impact Craters" http://www.psi.edu/explorecraters/ includes virtual tours of craters, a list of resources, and a site for submitting questions about impacts. A more teacher-oriented site is "Finding Impact Craters" http://craters.gsfc.nasa.gov/index.htm.

The Deep Impact mission, which observed the collision between a projectile and Comet Tempel 1, provides background material (including movies of impact experiments) at http://solarsystem.nasa.gov/deepimpact/science/cratering.cfm

Objects that burn up in our atmosphere and create meteors survive to strike the surface of the Moon. Observations of such impacts are discussed at http://science.nasa.gov/headlines/y2008/21may_100explosions.htm..

An online tool to demonstrate the way that crater sizes vary with impact angle, target and projectile material, and projectile speed is found at http://www.lpl.arizona.edu/impacteffects/, which allows visitors to enter desired values and see the size of the resulting crater.

3

Inner Fire: Volcanic Processes

INTRODUCTION

Volcanic activity provides some of the most awesome, dangerous, and beautiful natural spectacles on Earth. Such events can range from relatively placid, like the slow movement of lava from Kilauea's flanks to the sea, or sudden and dramatic, like the explosion of the summit of Mount St. Helens. In fact, these two sorts of events may seem to have little in common but the underlying mechanisms that drive them are rooted in similar processes.

At its simplest, volcanic activity can happen whenever some sort of melted or liquid material moves across or erupts through a surface of solid material. The liquid can find its way to the surface in many different ways. It can cause a quiet flow, or spray out in a plume. If it has trapped gasses within it, or is backed by pressure, it can also erupt explosively.

Volcanic activity has played, and continues to play, one of the most important roles in shaping the surfaces of the terrestrial planets. Along with impact cratering, volcanism is responsible for most of the features we see on Mercury, Venus, and the Moon today. Earth, with its highly active **plate tectonics**, has wiped out most of its history of both cratering and volcanism, but they remain key processes shaping the planet.

On all these worlds, volcanism is largely a case of liquid rock erupting through and then moving over solid, cooled rock. But this need not be the case, and is not the case, in the outer solar system. There some of the icy worlds and moons have been subject to "cryovolcanism," that is solid *icy* materials, like water and methane, being covered over by melted icy materials.

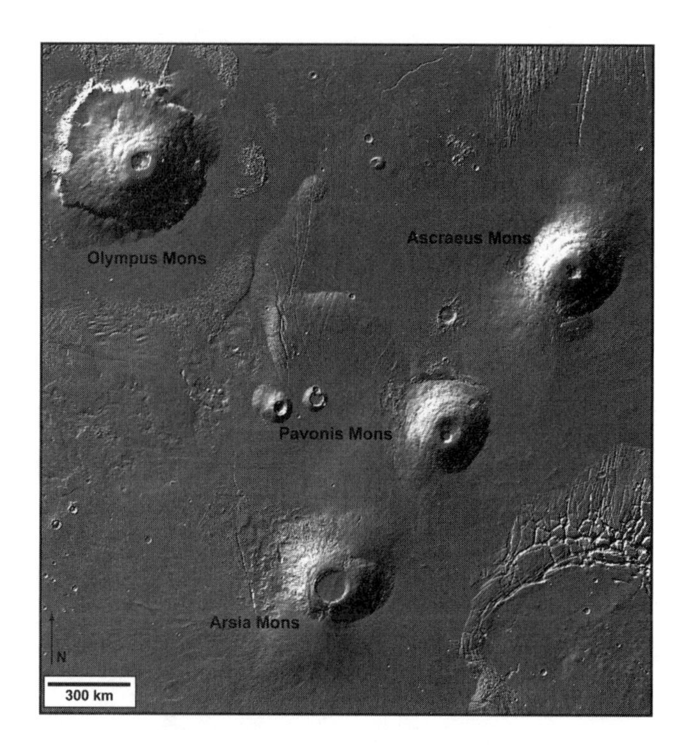

Figure 3.1 Volcanoes have been observed from space on several planets. On the top is seen the "Tharsis Volcanoes" of Mars, including Olympus Mons, the largest volcano in the solar system. On the bottom is a view of Mount Cleveland in the Aleutian Islands, as seen from the International Space Station.

PLANETARY HEAT BUDGETS

Heating Up—Sources

Having melted material on or within a planet prompts the question of what melted it in the first place. Melting requires heat. As it turns out, there are several key sources of heat energy that have been or still are available to the rocky inner planets. These sources are responsible for past and present volcanism.

- **Residual Heat of Formation.** From basic physics we know that gravitational (potential) energy is released when an object moves closer to the center of gravity of a body. The simple act of falling releases energy. If that energy can be absorbed, then it can be converted into heat. Because of this, the planets were heated by the act of formation. As small objects within the solar nebula (**planetesimals**) began to collect (**accrete**) into larger bodies, they released their potential energy, and heated up the newly forming planets.
- **Radioactive Decay.** Some of the material that accreted to form the inner planets was naturally radioactive. As these radioactive materials (**radionuclides**) decayed over time, they released energy, which was absorbed by and heated the surrounding matter.
- **Heat of Differentiation.** The material that collected to form the planets was not all of the same composition. Some of the material was lower density like water ice, some was of an intermediate density like rocky feldspar and quartz, and there were also high density materials like metallic nickel and iron. Originally all of these materials were distributed more or less evenly (homogeneously) throughout the new planet. However, denser materials are not stable on top of lighter materials. An example of this is a mixture of oil and water, as in salad dressing. The lower density oil floats on top of the higher density water. If you shake the bottle, mixing the two liquids, they will quickly separate themselves again based on their density, a process called **differentiation.** If you added peanuts to your salad dressing, they would immediately fall to the bottom, pushing oil and water out of their way on the way down. In fact, as the peanuts fall through the lower density materials, they release potential energy and generate heat. The same process occurred within the newly formed inner planets. Higher density metal material moved toward the center of the body, while lower density rocky materials moved toward the surface. This differentiation of materials released a great deal of energy, heating the interior of the planets.
- **Tidal Heating.** The word "tides" brings up images of sea level near the shore rising and falling over time. The gravity of the Sun and Moon are responsible for this phenomenon. As the Earth rotates under the Moon, and revolves around the Sun, the gravity of those bodies tugs on the Earth. Observations of the liquid water at the surface of the Earth show this very clearly. But the whole of the Earth, not just the sea, is being tugged by the

The Cooling and Aging Earth

In the mid-1800s, the scientific debate over the age of the Earth was spirited. Scientists realized that the heat produced by the Earth could provide a clue. Scottish physicist William Thomson performed a set of calculations starting with the assumption the Earth began completely molten. If it started at a uniform temperature of roughly 2000°C, the surface would cool rapidly by radiation, but the interior would cool much more slowly. How long would it take to cool down to the values seen today?

Thomson's calculations showed that the Earth was something like a few tens of millions of years old, perhaps up to a few hundred million years. He also provided supporting calculations using an independent technique showing the Sun was about the same age. While not nearly as old as biologists believed, Thomson showed that from a scientific point of view the Earth needed to be much other than a few thousand years.

Today we know Thomson's calculations were off by a factor of roughly a hundred: the Earth is 4.5 billion years old. Where did he go wrong? Thomson's calculations were made before radioactivity and fusion were discovered, both of which provide additional heat to the Earth and Sun. Without radioactive heating, the Earth would have cooled off long ago, as Thomson had calculated. For this work as well as excellence in a number of scientific fields, Thomson was rewarded with a title: Lord Kelvin. And it is in this way that Thomson's name lives on most famously: as the namesake of the Kelvin, the unit of temperature used by scientists.

gravity of the Sun and Moon, which causes tides inside the planet as well as at the surface. The deformation and friction caused by these tidal forces is a source of heat for the interior of the planet. Any planet can experience tidal heating from the gravity of another body. The efficiency of that heating depends on factors such as how close the bodies are, the eccentricities of the orbits, and more. The volcanically active moon of Jupiter known as Io suffers from massive amounts of tidal heating as Jupiter and its other moons pull at the small body.

- **Impact Heating**. Early in planetary formation, planetesimals of all sizes collected together to form **protoplanets**, which contained most of the mass found in the inner planets today. These very young planets suffered from collisions with more planetsimals, and even with other protoplanets. Some of these collisions were dramatic enough to shatter both protoplanets involved, eventually allowing for another protoplanet to arise from the collection of the shattered material. In other cases, the collisions were not so dramatic, but still dynamic enough to cause large explosions on the surfaces of the protoplanets. These explosions are impact events, and they were, and are, a source of heat for the surfaces of planets. Larger impacts can deposit a great deal of heat deeper in a planet, while smaller ones will only heat the surface.

Cooling Down—Sinks

Obviously, if there are sources of heat, there must be sinks for heat, that is, ways in which objects lose heat or cool off over time. Otherwise, everything in the Solar System would have melted and then vaporized completely. Instead, there are ways in which planets lose heat and cool down as time passes.

- **Convection.** Hotter materials are usually at a lower density than cooler materials of the same composition. As noted before, lower density materials will naturally tend to move above higher density materials. This phenomenon can be responsible for **convection** within the interior of a planet. Convection is a means of transferring heat by the actual movement of matter from one place to another. As material is heated near the hot core of a planet, its density goes down. It pushes upward against cooler material, rising toward the surface. As it rises, it also cools, releasing most of its heat near the surface, which eventually finds its way out of the planet completely and into space. The cool material then drops again, being pushed aside by newly heated material rising from the core. After descent, it can again be reheated, with its density again dropping. This cycle continues over and over, creating **convection cells** that effectively carry heat from the inside of a planet to the outside, cooling the planet. (See the chapter Winds of Change for more information on convection within atmospheres.)
- **Conduction.** Heat can also be moved within matter itself. For example, if one end of a spoon is held in a candle flame, the other end will slowly get warmer until it is too hot to hold, even though it never touched the flame itself. Heat was conducted within the matter of the spoon. In the case of conduction, the atoms within the solid material bump against one another, transferring energy. The atoms in the hot end of the spoon are moving faster, having more collisions with nearby atoms, making them move faster and having more collisions, as well. Eventually energy is transferred as heat along the length of the spoon. Such heat transfer also takes place inside of planetary interiors, with heat from the deep cores transferring outward toward the cooler surfaces.
- **Radiation.** "Radiation" can be a confusing word, since it is used with completely different meanings in science. It can specifically refer to the decay of naturally **radioactive elements**, it can refer to the particles emitted by radioactive decay, and is also used to refer to electromagnetic radiation (light), which is something completely different. In the case of heat transfer, **radiation** refers to the movement of energy directly through space via light. Anything not at absolute zero temperature has some energy within the atoms themselves. Light can be emitted from any such atom as the electrons circling the nucleus change their orbits. Assuming no other competing factors, a planet can be treated as a "black body," something that both emits and absorbs light energy. The energy emitted from a black body is described by a simple expression called the Stefan-Boltzmann law: $E = .\sigma T$. Where E is enertgy in watts per square meter, and T is temperature in Kelvin. The symbol σ is a constant (the Stefan-Boltzmann constant, 5.67×10^{-8} Watts m^{-2} K^{-4}). Atoms at the surface of any planet (or asteroid, comet, etc.) have been heated by energy from the interior, as well as by sunlight. Some of this energy is radiated, lost directly to space, based on the expression above.

More Cooling Than Heating

Currently the inner planets are operating with a deficit in their **heat** budgets. They are losing more heat from their interiors than their sources can replenish. Some bodies have solidified nearly completely, and are now too cold to fuel any significant volcanism or tectonic movement on their surfaces. Other planets, like the Earth, still have enough heat trapped inside to drive important processes, like volcanism.

Early in their histories, the planets were quite hot. As described above, they began by gaining heat from the infalling planetesimals that collected to form the bulk of their mass. Radioactive decay of short-lived radionuclides generated a pulse of heat that along with initial impact heating, provided the heat necessary to melt part of the interior of the inner planets. Once some melting occurred, the planets began to differentiate, with lower density materials moving toward the surface, and denser metals sinking to the core. Differentiation released even more heat, effectively melting the interiors of the planets until they were completely differentiated into light rocky crusts near the surface, a mantle of intermediate density, and high density core material.

Since that time, impact events have become less frequent. Although they can cause localized heating and melting of rocks, they are not a major source of heat for a planet. Radioactive materials do not all decay at the same rate, so initially after planetary formation there was a great deal of heat created from the decay of short-lived radioactive elements. After 4.5 billion years, there are still radioactive elements decaying within the planets, providing a source of heat, but not nearly as much as early in the life of the planets. Tidal forces allow for some small additional heating of the interiors of the inner planets, but this is a very minor source.

The result is that while all the inner planets had widespread active volcanism early in their histories, only the Earth still experiences it on a large scale. Why the Earth? The most likely explanation is a simple one—that size matters a great deal in heat retention.

Heat is *lost* through the *surface area* of a planet. Looking at the Stefan-Boltzmann expression, for example, we see that radiated energy is emitted as a function of watts per each square meter. But energy is *retained* within the entire *volume* of a planet, so heat is stored per meters cubed. The planet that will be the most successful in retaining its heat will be the one with the least surface area as compared to its volume.

This comparison can be expressed as the "Surface Area to Volume Ratio" of a planet. From basic geometry we know both the surface area and volume of a sphere:

Surface Area = $4 \pi r^2$
Volume = $4/3 \pi r^3$

So the Surface Area to Volume Ratio for a given planet of radius "r" is:

Ratio = $4 \pi r^2 / 4/3 \pi r^3 = 3/r$

Therefore a planet with a radius of 300 km has a surface area to volume ratio of 0.01, while a larger planet with a radius of 3000 km will have a smaller ratio, only 0.001. Obviously the larger planet will have more success retaining its heat over time.

The Earth is the largest inner planet, and so has been able to trap more heat, and for longer, than any of the others. Earth's small Moon has vast volcanic plains that are 3–3.5 billion years old. It may have had some small amount of volcanism as recently as 2 billion years ago, but very likely has been inactive since. Similar timeframes are likely on Mercury. Mars, the next largest in size, has giant **shield volcanoes** built from 1 to 2 billion years ago. There are some smaller flows that may date from only 10 to 200 million years old, but the planet is otherwise completely volcanically quiet. The story for Venus is not so simple. Recent research seems to indicate that the surface of Venus may still be undergoing some amount of volcanism, but nothing like the massive pulses of volcanism that nearly resurfaced the entire planet between 200 and 700 million years ago. The Earth, the largest of the lot, still sees dramatic volcanic activity. In addition, the internal heat of the planet drives vast tectonic activity, such as plate tectonics, which rapidly wipe out evidence for just how much volcanic activity is still occurring.

VOLCANISM IN ACTION

Moving Magma

Envision a large rocky planet that has undergone differentiation. Its interior has split chemically into three regions; a thin "crust" of a surface formed of low density rocks such as feldspar and quartz, a thick, deep "mantle" of moderate density rock such as olivine, and finally a central core made mostly of high density iron and nickel metals. The interior is mostly solid rock, not liquid, even though it is hot. This is because the pressure within the planet is very high, and although the rock is hot most of it cannot melt at high pressures.

The pressures and temperatures are high enough that the lower mantle rocks can slowly be deformed without breaking. They are not "brittle" but instead are "ductile." In fact, this solid rock is slowly convecting, bringing heat directly to the upper mantle and lower crust regions. The pressure toward the surface is lower, and the heat within the rocks is then sufficient to melt it in some areas forming liquid **magma**. Magma is simply the word for melted liquid rock found beneath the surface; when it erupts onto the surface it is called **lava**.

Magma gathers or pools within the solid rock into regions called **magma chambers**. The magma from these chambers can slowly work its way up through cracks in the brittle rocks above, or can be forced up more dynamically by a variety of processes. As noted, a liquid of a given composition is generally less dense than the solid of that composition, so the magma will naturally rise where possible. Magma chambers can be compressed as the solid rock around them is pushed inward by continued deformation. Magma may contain other volatile gasses trapped or dissolved within it, like water vapor, carbon dioxide, or sulfur dioxide. These gasses can provide additional pressure, forcing the magma upward. Eventually, through a combination of processes, liquid magma can erupt onto the surface as lava. Alternately it may also remain trapped underground and cool there as an intrusion.

Exactly how and where magma erupts, the nature and duration of the eruption, and then the landforms that result, are often a function of the chemistry of the magma, as well as relationships between volcanic and tectonic activity unique to a given planet.

Composition

The composition of magmas and their resulting lava is very complex, but there are some basic generalizations that are key to understanding volcanism on the inner planets. The terms *basaltic* and *silicic* refer to the composition of the melted rock itself. A basic question to be asked of any lava flow and resulting landform is if the flow is more basaltic or more silicic in composition.

Silicic lava, also called **felsic**, is largely composed of **minerals** like quartz (SiO_2), and some feldspars (solid solutions between $NaAlSi_3O_8$, and $KAlSi_3O_8$, and some $CaAl_2Si_2O_8$). This means silicic/felsic lavas are rich in silicon, aluminum, potassium, and sodium. These rocks melt at relatively low temperatures, as low as 650°C, so silicic lavas can reach the surface of a planet and still be relatively cool. Because of their composition and lower temperature, these lavas have a higher viscosity, and move more slowly.

Basaltic lava, also called **mafic**, is largely composed of minerals like olivine ($(Mg,Fe)_2SiO_4$) and pyroxenes (solid solutions between $Mg_2Si_2O_6$, $Fe_2Si_2O_6$, and $(Mg,Fe)CaSi_2O_6$). This means basaltic/mafic lavas are rich in iron and magnesium. Higher temperatures are required to melt these rocks: 950°C and above, and so basaltic lavas tend to reach the surface at higher temperatures. Basaltic lavas tend to have lower viscosities, and move more rapidly.

These are the extreme cases for lavas. There are lavas of intermediate composition, sometimes called **andesitic** lavas, with intermediate melting temperatures and viscosities. It is also possible for magma to change compositions as lava is released (or **extruded**). Some mineral compositions are very sensitive to temperature, and as a result the detailed composition of early lavas may not match later lavas extruded from the same magma chamber.

Magmas and their resultant lavas contain differing amounts of trapped gasses. More volatile substances, such as water, carbon dioxide, and sulfur dioxide, can be dissolved into or trapped inside magmas. As the magmas rise to the surface, and the pressure on them decreases, the gasses within them are released from solution, and become mobilized. This can mean that when the magma reaches the surface, it erupts explosively, as the gasses expand suddenly and dramatically.

Viscosity Predicts Behavior

How lava behaves, that is, how quickly and dramatically it erupts, if it flows as largely liquid or thick solid material, how fast it moves, and the landforms that it ultimately creates, is closely related to the **viscosity** of the lava. As noted, viscosity is determined by several factors including composition, temperature, and amount of dissolved volatiles.

Lavas with low viscosity tend to behave "nicely." They are not the trouble-makers of the volcanic world. Lower viscosity lavas erupt more quietly, releasing gasses easily as they come out of solution. These lavas flow more quickly, often in liquid form, and while they remain hot they can create rivers, lava channels, lava tubes, and collect into large lava lakes. The landforms they create are generally lower profile. Imagine a flow of melted chocolate across a table. When it is hot, it flows easily, and thins out over the table. When it cools, it has formed a very thin layer of hard chocolate, smooth and with little topography. You would not try to make a steep sided mountain out of liquid chocolate.

Lavas with high viscosity behave more dramatically. These lavas are the tempestuous members of the volcanic world. High viscosity lavas erupt more dramatically, sometimes in huge explosions. They do not release gasses easily. These eruptions may not contain much liquid at all, but instead be massive flows of mixed solid rock, ash, expanding volatiles, and clots of melted material that are called a **pyroclastic** flow. The landforms created by higher viscosity lavas can be much higher in profile. Imagine a flow of mashed potatoes, perhaps with chunks of potato still entrained within it. It does not move easily, and more potatoes can be poured on before the pile pushes outward and continues to flow over the table. A much steeper mountain can be made from a single "flow" of mashed potatoes than from melted chocolate.

Some lavas, like those that behave like mashed potatoes, are not typical fluids at all. A typical fluid (like most liquids we encounter every day) is called a **Newtonian fluid**, and knowledge of a Newtonian fluid's viscosity allows its behavior to be predicted. But mashed potatoes (or ketchup or paint), and some silicic lavas and pyroclastic flows, are not Newtonian fluids, but instead are **Bingham fluids** (or Bingham plastics). This means they will flow, but only after a certain force (a particular amount of shear stress) is applied to them. See the chapter Breaking Point, to learn more about stress, strain, and Bingham Fluids.

The difference in viscosity can be seen in different types of basaltic lavas. The two major forms of basaltic lava on Earth are **pahoehoe** and **a'a**, as the Hawaiian names have been adopted by the scientific community. Pahoehoe is relatively smooth and occurs with low viscosity. At higher viscosities a'a forms as a much blockier, rugged flow. Both forms can be seen in a single flow if viscosity changes or lava encounters steeper or shallower slopes.

VOLCANIC FORMS

Volcanoes

For most people, the word "volcanism" means "volcano." The two are not synonymous, but certainly mountains with lava and gasses coming from the top or sides are commonplace on Earth. However, a volcano is only one type of structure that can be built by volcanic processes.

Typically, volcanoes are masses of cooled lava or pyroclastic materials that have built up over time. A volcano may start forming around an opening or crack in the surface where lava, gas, ash and other materials are first extruded. This is the place where magma from an adjacent magma chamber has found access to the surface. Eventually, enough materials accumulate to form a roughly cone shaped structure that can range in size from a small hill to some of the largest mountains on the planet (or indeed, on other planets).

The nature of a volcano is related to the viscosity and composition of the materials that form it. Highly fluid, basaltic flows will pour from vents relatively quietly. The lava will flow outward, cool, and then be topped by another thin, wide flow. Over time, a very broad, flat mountain will build up, with eruptions continuing on the sides (flanks) of the volcano where additional lava breaks through. These are referred to as **shield volcanoes**. Examples of this sort of volcano are Olympus Mons on Mars, and Mauna Kea on Earth. In fact, suspected shield volcanoes in various sizes have been identified on all the inner planets.

More silicic materials will produce a much different volcano. Such lavas, ash, and pyroclastic flows will erupt from a vent with great energy, spewing forth with violence and cooling relatively quickly into chunky, thick layers. These will be topped by additional flows and blankets of ash. The resulting structure will be a very steep mountain, such as Mt. St. Helen's on Earth. Such volcanoes are called composite or **stratovolcanoes**, and Earth is the only planet where they are found.

Figure 3.2 In contrast with the (relatively) gentle eruptions found at more basaltic (shield) volcanoes, volcanoes with more silicic lavas (stratovolcanoes) tend to experience more violent and destructive eruptions. Mount St. Helens, seen here during its 1980 eruption, lost most of its summit in an explosion that showered the Western United States with ash.

Flood Basalts

While the word "volcanism" inevitably brings up images of steep, conical shaped volcanoes, they are not the most dominant volcanic form in the inner solar system. In fact, the most common volcanic forms are wide expanses of cooled, hardened lava known as flood basalts or volcanic flood plains. These are often massive and formed in flat layers over time. All the inner planets have wide expanses of flood basalts.

Flood basalts are much, much bigger than a typical lava flow from the flanks of a volcano, are vastly more massive than the volume of the volcano itself. Flood basalts are unmatched as a volcanic form in terms of size. A single eruption of flood basalts can be thousands of cubic kilometers in volume, and cover over a hundred thousand square kilometers as it spreads out. Furthermore, most flood basalt plains are formed from many eruptions layered on one another. The Siberian Traps flood basalts on Earth cover more than 1.5 million square kilometers today, and may have been several times larger when originally formed.

The extent of such incredible formations can be seen by looking at the full Moon. The dark areas are flood basalts that pooled into the bottoms of huge impact basins. Mercury and Mars have wide volcanic plains of flood basalts. There are also large expanses of flood basalts on the Earth, and there were certainly more in the past, now wiped out by processes such as erosion and plate tectonics.

Other Volcanic Forms

While volcanoes and flood basalts are the most dramatic signs of volcanism, volcanic activity can also produce other landforms. Some of these are widespread, for instance **lava tubes**. Lava tubes are the pathways taken by lava near the surface, now drained of lava. They often are similar to caves, though they can also be open if their roofs have collapsed. Lava tubes of various sizes have been seen on many of the terrestrial planets. **Cinder cones** are formed by the collection of glassy solid fragments rather than successive lava flows. These fragments, called cinders, form as a spray of tiny magma droplets cool in the air. Some volcanic forms are peculiar to specific places or unusual circumstances: for instance maar craters, caused when magma encounters water near the surface, leading to a huge steam explosion, and pancake domes, endemic to Venus and thought to depend on that planet's thick atmosphere.

CONCLUSION

Volcanic activity is one of the most important processes shaping the surfaces of the inner planets. Forms such as flood basalts cover vast portions of the planets' surfaces, and are the canvas for much of the activity that came afterward, such as the formation of impact craters, river channels, tectonic faults, and more. The power for all this volcanism lies in several key heat sources, some of which were more important in the past, and some that still provide heat to drive volcanism today. The Earth, as the largest inner planet, has done a better job at retaining its heat any other, and so has more active and more recent volcanism than the other terrestrial planets. As the planets continue to cool, active volcanism will end. Even the Earth will someday have its volcanoes permanently become quiet. But like the massive Olympus Mons on Mars, the largest volcano in the solar system, the evidence for the Earth's highly volcanic past will remain for billions of years.

FOR MORE INFORMATION

The USGS Volcanoes Hazard Program is at http://volcanoes.usgs.gov/, including a fine photo glossary.

The Smithsonian Institution also has a set of volcano information, at http://www.volcano.si.edu/, including volcanoes outside of the United States.

Oregon State hosts the Volcano World Web site, which features extraterrestrial volcanoes (http://volcano.oregonstate.edu/volcanoes/planet_volcano/) including an extensive set of pages centering on Venus (http://volcano.oregonstate.edu/volcanoes/planet_volcano/venus/intro.html).

An overview of lunar volcanism can be found at http://www.geology.sdsu.edu/how_volcanoes_work/moon.html as part of another overview of overall volcanism (http://www.geology.sdsu.edu/how_volcanoes_work/index.html).

4

Breaking Point: Tectonic Processes

INTRODUCTION

Planetary surfaces are constantly under the influence of internal forces. This generally becomes obvious only when a "failure" occurs. In human interactions when there is failure, we might think it is somebody's fault. For rocks, however, things are reversed: a **fault** occurs where there is **failure**. Geologists define a fault as a crack along which rocks move with respect to one another. The San Andreas fault in Southern California is a famous example. The Earth's internal stresses are always pushing on the fault zone, and every once in a while, the stresses reach a point that overcomes the strength within and the friction between the rocks, and there is motion along the fault. The result is an earthquake, occasionally a very dramatic one. And earthquakes can result in damage to highways and buildings, the initiation of landslides, and even the formation of giant tidal waves called **tsunami**.

If people have heard the word **tectonics**, it is probably in the context of the phrase **plate tectonics**. Plate tectonics (see the chapter about Earth) describes a particular kind of tectonic activity, but tectonic processes as a whole cover a much broader spectrum of processes and landform creation on planets. In its simplest form, tectonic activity is what results when a planet's internal forces disrupt its surface. This disruption can create a host of events and landforms, including fault zones, mountains, gorges, landslides, and earthquakes.

Evidence for tectonic activity can be found on all of the inner planets. Any planet with a **brittle** (breakable) crust is going to show some signs of wear and tear. How and why these features formed is critical to our understanding of the evolution of all planetary surfaces.

MOVING, BENDING, AND BREAKING

An understanding of tectonics is built on knowing what puts stress on planetary rocks and how the rocks will eventually move, bend, and break when overstressed. Forces such as gravity can push or pull on rocks and have to overcome other forces, such as the frictional force, to get rocks to move. Forces must overcome the innate strength of the rocks themselves in order to get rocks to bend and possibly break. The water content of rocks, their temperatures, the pressures they are subjected to, and if they are whole, riddled with fractures, or even in pieces are all factors in how the rocks act under stress.

Gravity

Gravity is a fundamental force that puts stress on planetary rocks, as well as governs the motions of some rocks after a major event like a landslide. Tidal forces push and pull rocks within a planet, heating the interior, but also put stress on planetary rocks that compresses them in some places and pulls them apart in others.

The most straightforward models of how gravity stresses rocks are found in basic mechanical physics. Imagine a simple case where gravity is acting to force a block downward against a table like the left panel of Figure 4.1. From basic physics we know that an object sitting motionless on a table is always subject to forces, even if it does not look like it. All forces are resisted by equal and opposite forces, so the force balance for a motionless block looks something like:

The force of gravity, F_g is:

$$1 \qquad F_g = GMm/R^2$$

Where G is the gravitational constant, M is the mass of the planet we are on, m is the mass of the block, and R is the distance of the block to the center of the planet.

The resultant force, F_r, is found in the strength of the wood of the table resisting being crushed by the force of gravity on the block. If the block is not very heavy, and the table is sturdy, then the table is not crushed and instead supports the block. If the block were much more massive, or if the gravity of the planet were magically increased (or if the same block and same table were transported to a planet with increased gravity), then perhaps the table would not have the strength to support the block, and would collapse.

Friction

As objects move against one another, they resist motion due to frictional forces between them. The nature of this resistance is affected by factors such as the presence of water or small particles, the smoothness or roughness of the surfaces, etc. Imagine we place the same block on an inclined wooden plank (Figure 4.1, right panel), which means a component of the force of gravity will try to get the wood block to move down the plank. Gravity still pulls directly downward (F_g), as we have come to expect, but the force can be split into two components. Part of the force is pulling directly into the incline (F_n, the **normal force** component) and therefore pushing the block against the wood beneath it. And the other component of the force of gravity is pulling straight down the slope (F_L) trying to get the block to slide.

If we assume the force of gravity is not crushing the wooden plank, then the resultant force F_r once again perfectly balances the force in the opposite direction, F_n. But there is still a component of the force of gravity trying to get the block to slide, F_L. Imagine our wooden plank and block as starting flat on the ground, and then having one end slowly lifted. We know that we can lift the end of the plank up just a little from the ground, and the block will not slide. In fact, we can lift it quite a bit before the block slides down. Of course what is happening is that friction between the block and the plank is resisting the force of gravity. The friction between two objects is described by the coefficient of friction, μ. So the force balance up and down the plank, right before the block slides, is:

$$F_r = F_N \quad (2)$$
$$F_f = F_L$$
$$F_f = \mu F_n \quad (3)$$

As the angle of the incline is increased, more force is exerted straight down the slope. For the block to move, gravity must overcome the force of static friction between the block and the plank. When the angle is high enough, there is an instant right before the block begins to move. At that moment, we can use a little trigonometry and find:

$$F_f = \mu F_n = \mu F_g \cos\theta \quad (4)$$
$$F_L = F_g \sin\theta \quad (5)$$
$$F_f/F_L = \mu\ F_g \cos\theta\ /\ F_g \sin\theta = \mu(\cos\theta/\sin\theta) \quad (6)$$

When F_f and F_l are equal, then F_g cancels out. This has far-reaching consequences for understanding how rocks move on other planets. With all other factors being equal (no wind, water, glue, and such) this block and plank will behave exactly the same no matter what planet they are on. The fact that Mars has a lower surface gravity than Earth, for example, is not a factor. Gravity is literally not in the equation when it comes to the moment an object begins to slide downhill.

One of the practical results of gravity canceling out of this equation is that piles of loose, dry material found on any planet will form much the same slope on their sides. When a pile is formed, say by wind piling up sand, the sides of the pile will become steeper and steeper as more material is added until it reaches a critical slope. This slope is called the **angle of repose**. The angle of repose is somewhat different depending on the nature of the particles in the pile, but it generally varies between 25 and 40 degrees. Dry sand, for example, has an angle of repose of about 33 degrees no matter what planet it is found on. Once the sides of a pile have reached the angle of repose, any further material added will fall down the sides and probably dislodge other particles as they fall. More and more sand can be added, and the pile will grow taller, but the angle of the slope on the sides will remain the same.

Stress and Strain

Another example of forces acting on an object is a person pushing on a shopping cart. You push on the shopping cart, and the cart moves. You are confident that a force was exerted since you can see the cart rolling away, whereas it used to be stationary. Of course, if you push on a cart and the cart does not move (because the wheels are clogged with gum) you are still exerting a force. You know that is true because you can feel the effort of pushing. What is less obvious is that the cart is pushing back, just like the table was pushing back on the wooden block.

Pushing on a cart that will not move is like trying to squash the cart. The internal strength and forces within the metal itself resist being squashed, and push back at you. The harder you push, the harder the metal pushes back. Nothing is moving, but there are forces acting within your arms and within the cart.

The end result, naturally, is that after a bit of pushing you will choose a different cart, since this one is clearly not going to do the job. In our hypothetical case, though, the end result of pushing harder and harder is that something eventually gives way. Either your arms give out, or the cart is bent or broken.

Scientists refer to this force of pushing as the **stress** on an object. And like highly overstressed people, highly overstressed objects eventually suffer some kind of breakdown. This breakdown is the **strain** on the object—how much movement, bending, or other deformation is caused by that stress force. In our case with the cart, you might imagine that you push on the cart until the metal bends. The bending is the strain suffered as a result of the stress you applied.

Another case we can imagine is a building. Imagine a large brick structure built on top of a rock foundation. The force balance is between the weight of the brick building pushing down, and the strength of the stone pushing back. Imagine we add more and more stories to the building. It gets heavier and heavier—there is a reason you can't build a tower infinitely high. Eventually, either the brick on the bottom stories will crush, or the weight of the building will crack the stone foundation under it and actually break the ground. The moment before the ground breaks, there are great stresses in the building and the rock. The rock breaks, and suddenly the force is able to create deformation and motion. The forces were there all along but were not obvious until the rock cracked.

Technically, stress is a measure of the total of all the internal forces acting inside a body along imaginary surfaces within that body. Some can be from the top, others the bottom, some from the side, and some from within, like the upper stories of a building pushing down on the lower stories.

$$\sigma = F/A \tag{7}$$

Total stress, σ, is the average amount of force per unit area. It is equal to the force being applied over the area to which it is applied. This relation tells us that stress is related proportionally to force, but inversely to area. So the larger the area over which the force acts, the lower the overall stress. Some forces are small, but act over such small areas that they can still cause strain. The reverse is true for some larger forces, which might easily break rocks, but act over such large areas that the average amount of force is too low to strain the rocks.

Bending versus Breaking

Rocks do not immediately break as soon as a force is applied. We know from our own experience that some materials have more internal strength than others and take more force to break. We also know that there are materials that will bend instead of breaking, or at least will bend a little before they do break.

Materials that have a tendency to break are referred to by geologists as **brittle**, while those materials that are more likely to bend are called **ductile**. It may be hard to imagine a layer of rock bending instead of breaking. But one of the interesting things about rock is that it can act in either a brittle or ductile manner, depending on conditions. Cold rock under low pressure, as at the surface of the Earth, is generally brittle. But deeper inside the

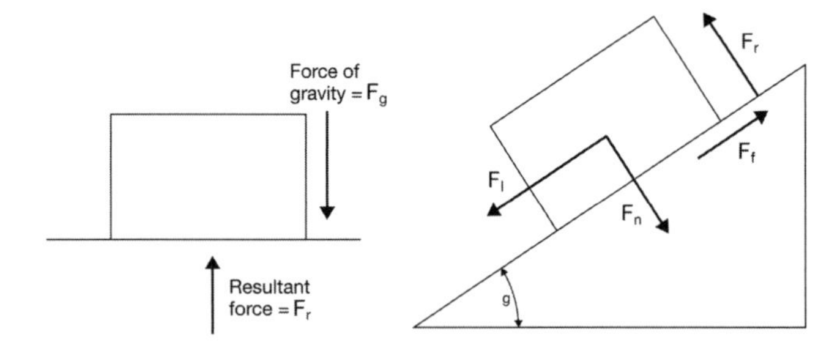

Figure 4.1 Simple diagram of block and stress force.

Earth, that same rock at higher pressures and temperatures will bend and deform, rather than break.

Some "bending" is permanent, and other bending is not. Imagine a wooden branch. If you bend it slightly then stop pushing, it springs back to its original shape. You applied a force (stress) and noted the bending (strain) but when the force was lifted, the evidence of the strain disappeared. This is referred to as **elastic** behavior, and many materials will behave this way if the stress is small. Now imagine a metal rod, or perhaps a coat hanger or paper clip. You can bend it slightly and have it spring back, but push on it too hard, and the bar will be permanently deformed. This is called **plastic** behavior. If you then push on the metal bar very hard, you may find that it will bend at first but then suddenly break. You have forced the bar to go from behaving in a ductile to a brittle fashion.

There are of course forces that can twist or shear rocks, not just push or pull on them. This can happen when the top of a rock is being pushed harder than the bottom or if the top and the bottom of a rock are being pushed in different directions. This **shear stress** is an important factor when we think about how some materials move, flow, and eventually fail.

A small mound of mashed potatoes on a table will not move. It will retain its shape until a force is applied. If you push down on the mound with your hand, the potatoes will start to flow outward and spread over the table. Mashed potatoes will also flow if you pile them high enough. Then the force of their own weight will be enough to cause them to flow outward and flatten until they have spread sufficiently. Then they will stop moving, having achieved another equilibrium, which you can upset by adding more potatoes to the top and watching the bottom flow outward again.

This is an example of a **Bingham fluid** or **Bingham plastic**. Within our everyday understanding, most fluids are **Newtonian fluids**. If you apply a force to these fluids, they immediately flow. The only parameter needed to really understand a Newtonian fluid is its **viscosity**. In contrast, a Bingham fluid will not flow until a certain stress force is applied, called the **yield stress**. To describe a Bingham fluid, one needs to know both the viscosity and the yield stress.

Examples of Bingham fluids in nature include certain kinds of: muddy debris flows, pyroclastic flows, landslides, and lava flows. In the case of lava, it is known that the presence of some solid crystals within the lava can cause it to behave as a Bingham plastic. Such a lava flow will often have a characteristic parabolic shape along the front of the flow because of the nature of the shearing along the model layers inside the flow itself. The muddy fluid-like ejecta around craters on Mars have also been modeled as Bingham fluids.

Compression and Extension

Most major landforms on planets are a result of forces acting in one of two ways. Compressional forces push rocks together. Rocks in **compression** can break, creating a fault where one side of the rock is pushed up and over the other. The faults created in rocks in compression are called **thrust faults**. **Extensional** forces pull rocks apart. Rocks under **tension** can also break, creating a fault where one side of the rock slides downward forming a **normal fault**.

The landforms created by compression include mountains, thrust faults, and overhanging ridges, and the landforms created by rocks under tension include troughs, extensional valleys called **graben** (*grah-ben*), and basins.

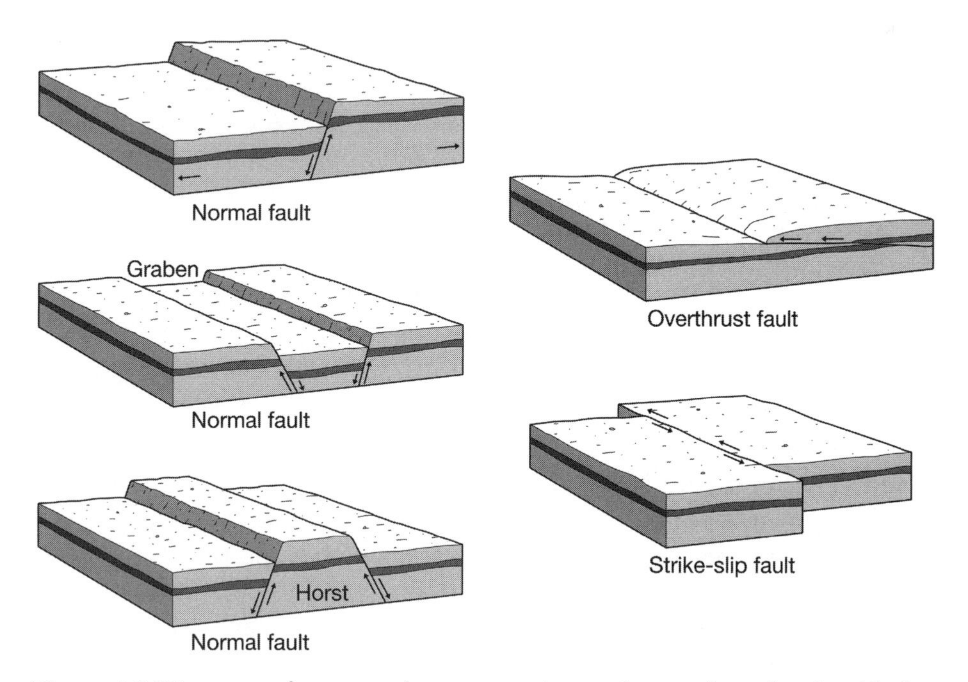

Figure 4.2 Diagrams of crust under compression and extension, showing blocks moving up and down, and close up of faults with movement arrows.

Mass Wasting

All of this bending, breaking, sliding, and flowing means that material in high places often finds its way to lower places by one means or another. **Mass wasting** is the movement of material down slope as a result of gravity and other contributing factors. A **slope failure** is a type of mass wasting where rocks move down slope after some kind of breakage or strain. Slope failure is typified by events such as landslides, rock falls, and debris slumps. Mass wasting can be sudden, violent, and dramatic, or slow and continuous.

The long-term force of gravity can be more than enough of an impetus for downslope movement, but sudden mass wasting events can often be triggered by something else. On bodies like the Moon and Mercury, small impacts can dislodge material on the rims of craters and allow for slumping to occur down crater walls. Wind erosion on Earth and Mars can scour the rocks on a cliff face, eventually leading to a landslide. On any planet, seismic waves from tectonic events can shake loose material already at the angle of repose to form a rock fall.

Water can play an important role in mass wasting processes. In addition to freeze–thaw cycles that serve to break down rocks and uplift soil, water can lead to **flows** of all kinds. Flows are another major type of mass wasting, and happen when surface materials, water, and possibly entrained volatiles like air or other gasses all mix together and move like a fluid down slope. Initially, a small amount of water mixed into a material can serve to strengthen it. The strong surface tension of water acts like a bonding agent. This can be seen easily in the difference between dry and wet sand. A pile of dry sand cannot hold a slope higher than the angle of repose, while moist sand can be made to form relatively sturdy vertical walls. But if the sand is completely drenched with water, the water will surround and isolate each sand particle. When that happens, the force of friction between the particles is lost, and the sand can no longer hold a slope at all.

A major rainfall can change the amount of water found in the pore spaces within the rocks of a cliff face, and suddenly allow part of the cliff to flow. On Earth, this can cause mud and debris flows that can be quite dangerous and damaging, often due to the fact that they are unexpected. Structures can be built in areas that appear solid, but a drenching rain can mobilize the underlying materials and lead to destructive flows. Tectonic events that occur underwater, such as a sudden underwater landslide, earthquake, or seismic waves generated from a volcanic eruption can have additional dramatic effects in the creation of a tsunami. Tsunami are massive waves in large bodies of water. Although not all tsunami can travel across an ocean, they can still cause major geologic changes in coastlines near the areas where they originate.

Some landslides, called "long runout" landslides, travel far greater distances than predicted given the vertical drop and the needed stresses.

It is suspected that such landslides were made temporarily more mobile by the process of **acoustic fluidization**, where acoustic waves within the media served to lower the forces needed to move the materials, creating a temporary fluid-like flow. Acoustic fluidization may also play a key role in the mobility of "weak" faults and the slumping of impact crater walls.

SEISMOLOGY

Any compression, extension, breaking, and sliding of crustal rocks is accompanied by waves of energy that propagate into the interior and along the surface of a planet. Just like a bell vibrates after being struck, a planet also vibrates during and after a tectonic event. The waves of energy within the bell cause it to shake rapidly, and thereby force the air between the bell and your eardrum to vibrate as well, producing sound. In the same manner, the waves of energy in a planet that propagate away from the location of a

Figure 4.3 This Magellan radar image of Venus shows graben criss-crossing the surface, indicating extension. NASA/JPL

tectonic event cause the ground to shake. This shaking is what we are familiar with as earthquakes. **Earthquakes** (or say, Moonquakes or Marsquakes) can range from the nearly undetectable, even to sensitive instruments, to the easily noticeable and quite damaging. You have probably heard of the **Richter Scale**, which was developed as a means to quantify the relative amount of shaking produced by a tectonic event. But you do not need a huge earthquake rating a nine on the Richter Scale to experience **seismic waves.** If a heavy box falls over in a nearby room, you can probably feel the floor shake. This is a tiny "housequake," where the tectonic event of the box falling causes small waves of energy to propagate through the house, shaking the floor.

Wave Basics

Tectonic events produce a complicated set of seismic waves that can interact with one another and with the material through which they are propagating. There are essentially two ways waves can travel around a planet—along the surface as **surface waves**, and through the planet's interior as **body waves**. An earthquake will create all of these wave forms, as well as **interference waves** where waves interact with one another or with changes in the media (e.g., the kind of rock, soil, etc.) through which they are propagating.

The body waves, the waves that move through a planet's interior, include **P-waves** (Pressure waves, also known as Primary waves), **S-waves** (Shear waves, also known as Secondary waves), and interference waves. Waves in a planet can be modeled by energy being transferred along a spring as an analogy. If the spring is pulled or pushed, then a back-and-forth wave moves down the spring. This is called a **longitudinal wave** (also **compression wave**), and P-waves are one example. If the end of the spring is moved side-to-side, then a side-to-side wave is created. These are **transverse waves**, and S-waves are one example of these.

There are two kinds of surface waves; **Rayleigh waves** (interacting P- and S-waves) and **Love waves** (essentially a polarized S-wave). Love waves produce a back-and-forth shaking motion. Rayleigh waves roll along the surface of the Earth like waves on the ocean, and so produce an up-and-down motion at the Earth's surface as well as a back-and-forth motion.

The Energy of Waves

Seismic waves do not all travel at the same speed. After a seismic event, the waves propagate outward with the P-waves fastest followed by the S-waves. The surface waves are much slower, with the Love waves and then the

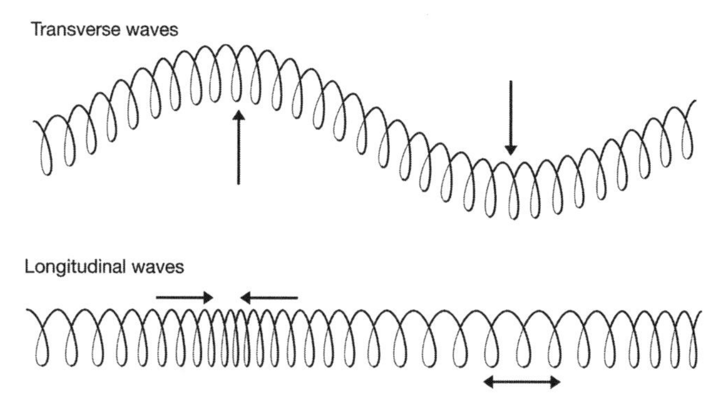

Figure 4.4 A common childhood toy is an excellent way of showing different wave types. Transverse waves result from shaking perpendicular to the direction of wave motion, while longitudinal waves result from shaking parallel to the direction of wave motion.

Rayleigh waves at the slowest speed. Of course, in this case, slow is relative. Seismic waves travel very quickly, on the order of kilometers per second. The speed of the waves can change depending on temperature, the nature of the rock, water content, density, pressure, and more.

Without providing a full derivation of the relationship between energy (E), frequency (f), density (ρ), amplitude (A) and speed (v) for waves, we'll instead simply state: $E = 2\pi^2 \rho f^2 vA$.

This equation has very important implications for how waves travel through different media, and more importantly, how the energy is partitioned between these different variables when a wave travels from one material into another.

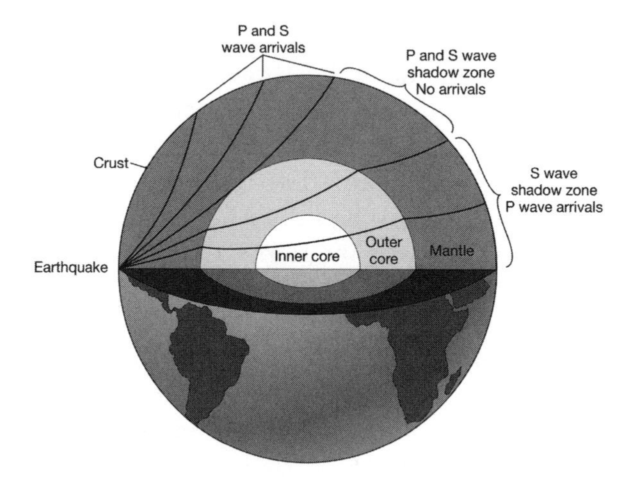

Figure 4.5 The differential speeds and properties of seismic waves allow the interior of the Earth to be probed. Some waves only travel through solids, which has been used as evidence that the Earth has a liquid core. The precise timings and locations of wave arrivals has also been used to measure the existence and properties of various layers in the Earth's interior.

Assume a solid rock suffers a widespread failure, and a large earthquake results. A wave of a given energy, frequency, speed, and amplitude propagates away from the **epicenter** of the earthquake through the solid rock. It passes by city "A" with little damage done, since the high density of the rock means that the amplitude of the wave, the factor that largely contributes to the amount of shaking, is relatively small. But then the rock layer ends and instead there is a layer of sediment. The wave continues through the ground, of course, but alters as the density of the new medium goes way down. For the wave to conserve energy, another variable has to increase to balance the decrease in the density from rock to sediment. This usually means the amplitude of the wave increases, which means more shaking. Now the wave passes by city "B" built on this sediment, and there the damage is extensive. Even though this city is farther away from the epicenter than city "A," the material the city was built upon had dramatic consequences for this, and any other subsequent earthquake.

This scenario is not fiction. In 1985, a large earthquake occurred about a hundred miles away from the city of Acapulco, in Mexico. Acapulco is largely built on coherent rock, and so shaking was held to a minimum. Almost no damage occurred in Acapulco. But the wave continued to travel through the ground, and another hundred miles further on it reached Mexico City. Mexico City is built on the site of an ancient lakebed, underlain with low-density sediment. Even though Mexico City was twice as far from the earthquake as Acapulco, the damage in Mexico City was vast. More than 5,000 people were killed.

Waves as Probes

Earthquakes can be terrible natural disasters. But they can also be exceptionally useful for the study of planets. The detection and interpretation of seismic waves is a critical scientific tool. Seismic waves can do what no instrument, probe, or drill can do—they can travel right through a planet and emerge out the other side. As they do, they change based on the structure of the planet's interior. Careful study of seismic waves gives us the opportunity to map the inside of a planet right down to the core.

For example, P-waves can move through both solid and liquid material, but S-waves can only move through solids. P-waves (pressure waves) are formed by pushing-pulling on material, and even liquids respond to pressure. However, S-waves (shear waves), require that the material be capable of transferring energy through shear in the form of a shear stress. Liquids do not have any cohesion, structure, or friction that allows them to transfer energy between layers, so S-waves cannot propagate through them. Because of this fact, we can use earthquakes to create a map of the liquid and solid places within the Earth. Knowing where the earthquakes originated and tracking the arrival times of P and S-waves around the world, we now know that the outer core of the Earth is in fact liquid. It will not allow S-waves to propagate through it. We also can find the locations of boundaries between different parts of the Earth's interior, since seismic waves can be partially reflected at these boundaries, not just travel through completely intact.

In addition to the arrival times for different waves, scientists can study how rapidly the waves die off after an earthquake. Seismic waves can vibrate through a planet long after the event that caused them is over. This is like the sound of an acoustic guitar after a string is plucked. The event, the plucking of a string, is short in duration but the waves that are responsible for the sound continue for some time. For guitars, the length of time the sound continues after the strings are touched (that is, the length of time the strings and sound cavity continue to vibrate) is called the "sustain." Guitars are designed with shapes and materials that encourage a lengthy sustain. The surfaces and interiors of planets are composed of any number of materials, all in different states. Some of these will be favorable to allowing energy waves to propagate, while others will absorb or damp the wave, forcing it to die off or **attenuate** more quickly.

During the *Apollo* project, several seismometers were placed on the Moon. Studies of the data from these instruments showed that seismic waves attenuated much less quickly on the Moon than on the Earth. Since both the presence of water as well as higher temperatures can attenuate seismic waves, it is likely that the Moon is much drier and cooler inside than the Earth. This is consistent with the chemistry of lunar rocks, as well as with the size of the Moon.

Paleoseismology

Seismology has been critical to our understanding of the Earth. With the exception of data provided by four lunar seismic stations placed by the *Apollo* astronauts (and turned off in 1977), seismic data has only been available for the Earth. Although a crude seismic instrument was built by the Chinese in the second century, reliable seismometers have only been available since the 1880s. However, resourceful geologists have been able to obtain useful information for earthquakes prior to the 1880s. Field work can show the relative motion of layers of sand, from which individual events can sometimes be reconstructed. In some cases, like that of the 1755 Lisbon earthquake, contemporary written records of damage in various cities and a knowledge of construction techniques can point to the most likely epicenter locations (which can then be investigated via field work). And in at least one case a giant earthquake in the Pacific Northwest of the United States was studied by looking at the growth rings of trees killed by the quake, with the event's exact timing determined by knowing the time of a tsunami arrival in Japan. Indeed, while no large earthquakes have occurred in the Washington–British Columbia region since the advent of seismometers, paleoseismic studies show that the area is at risk of another giant earthquake in the next few centuries.

CONCLUSIONS

Tectonic processes cover a wide range of activity on planetary surfaces including mountain building, earthquakes, landslides, and much more. Any time crustal rocks are broken and mobilized, tectonic processes are at work. Understanding the key forces of gravity and friction, the nature of stress and strain, and the basics of wave creation and propagation allows planetary scientists to model tectonic activity and predict the landforms that will result. The presence of water, among other factors, can have a major influence in the final outcome of the tectonic event. Rocks under stress can bend and break, causing seismic waves to propagate outward and through the ground of the planet. Such waves can provide scientists with a means to see into the interior of the body and map the internal structure. As a whole, tectonic processes represent some of the most common and widespread forms of activity seen on surfaces of the planets in the inner solar system.

FOR MORE INFORMATION

Graben formation is an important tectonic process, animated at the USGS Visual glossary (http://earthquake.usgs.gov/learning/glossary.php?term=graben), along with other geological terms.

A full discussion of earthquakes and how they are measured is provided by the Missouri Department of Natural Resources (http://www.dnr.mo.gov/geology/geosrv/geores/MeasuringEQs.htm), naturally including a discussion of the very large New Madrid earthquakes of the early nineteenth century.

Plate Tectonics: How It Works by Cox and Hart (1986, Blackwell Scientific Publications, Palo Alto) is an excellent introduction to plate tectonics, a critically important process on Earth.

Geodynamics (Turcotte and Schubert, Cambridge University Press, New York, 2002) is a technical introduction to tectonics designed for more advanced students.

5

Winds of Change: Atmospheric Processes

INTRODUCTION

Weather is a human obsession. Anecdotally, the weather is the one topic of conversation supposedly fit for any gathering at any time. It changes constantly, day to day, hour to hour, and sometimes minute to minute. The weather can be a source of great enjoyment, as in a beautiful, pleasant day, or a source of tragedy, as in a destructive hurricane, and everything in between. So our obsession with the weather in our **atmosphere** makes sense—it impacts every aspect of how people live. Understanding and predicting the weather is important business, so much so that one can develop an idea that the Earth's atmosphere must be incredibly thick, extending far away from the planet.

Yet all weather on Earth happens in the lowest and most dense part of our atmosphere, called the troposphere. The thickness (height) of this layer varies, but generally does not exceed 16 km above the surface. This is only one-quarter of one percent of the Earth's radius (0.25 percent). If the Earth were the size of an apple, the troposphere would be less than the thickness of the apple's skin. All this weather that keeps us so occupied is found entirely in a narrow shell of gas around the planet.

But the troposphere is only one part of the atmosphere. Atmospheres can have many layers and all atmospheres have some kind of "structure," that is, aspects like composition and density that change with distance from the surface. Atmospheres do not have sharp, perfect boundaries when they reach outer space. Instead, they grow ever more tenuous with height, eventually becoming nothing more than a scatter of fast-moving molecules and atoms. Although the "top" of any atmosphere is not perfectly defined, for the Earth it is around 100 km above the planet's surface.

But the Earth's atmosphere is by no means typical of what we find in the solar system. In fact, there is no "typical" atmosphere. They are all different, ranging from the practically nonexistent, like the vanishingly thin

Figure 5.1 The potential power of the weather was seen in 2005, when a weather satellite caught this view of Hurricane Katrina, an intensely destructive hurricane. NASA/JPL/Cornell

gasses near the lunar surface, to something like that of Jupiter, a gas giant that could be considered more atmosphere than anything else. Each planet's atmosphere is unique, but all are dynamic, constantly moving and changing. Some changes happen very slowly, noticeable only over the entire life of the planet, and others happen very quickly, like what we see in the weather of our own planet Earth. And as these atmospheres change, they in turn create changes in the world around them, forming dunes, driving dust storms, altering **climate**, and possibly making a planet suitable (or unsuitable) for the presence of life.

THE DEFINITION OF "ATMOSPHERE"

If there is no such thing as a typical planetary atmosphere, then how is one defined? The answer is not perfectly clear. As with most phenomena in the universe, atmospheres refuse to fall into neat and tidy categories. Scientists describe the Moon as having an atmosphere. Most people would not really recognize the gasses around the Moon as being an atmosphere, since they are only familiar with the Earth's relatively thick and complex example of what an atmosphere might be. However, gas molecules and ions of hydrogen, helium, neon, and even potassium and sodium surround the Moon. This atmosphere is extremely thin, and constantly "leaking" away into space. Some of the lighter elements, like helium, for example, can escape from the surface within hours, while others might take months. The interactions of the Moon's surface rocks with the **solar wind** and with impacting dust and larger bodies constantly replenishes this ghostly thin layer of atoms and molecules.

On the other end of the spectrum we have Jupiter, a gas giant planet that is essentially atmosphere all the way down. There might be a rocky core in

Blue Sky on Mars?

It is well known, of course, that on Earth the sky is blue. The color of the sky on other planets is a different matter, however. While it might be possible to predict sky colors using theory alone, we have some information from space probes. The *Viking 1* lander returned images from the surface of Mars in July 1976 that, to the astonishment of many, indicated the Martian sky was blue. But more accurate calibrations of the images soon made it clear that the sky was not blue, but instead "salmon-colored." Since *Viking*, more spacecraft have visited Mars, carrying different types of cameras. The current consensus is that the typical daytime sky is "scarlet," due to the dust carried aloft in the atmosphere. As on Earth, the sky color changes near sunrise and sunset, though unlike Earth it becomes less red rather than more red during those times.

We have much less information about Venusian skies, derived from Soviet landers of the 1980s. While the thick cloud cover on Venus prevents the Sun from ever being seen on the surface, evidence suggests the sky color there is an orangey-red.

its center, but it is surrounded by dense layers of gas, mostly hydrogen and some helium, five times deeper than the diameter of the Earth. Again, using the Earth's atmosphere as an example does not help much as a model here. While the top of the atmosphere on Earth may be hard to point to, the bottom is easy to spot. The bottom of the Earth's atmosphere clearly stops where something like the ground or the ocean starts—that is, where the gas of the atmosphere hits an obvious boundary with a solid or a liquid. Jupiter's atmosphere simply gets denser and denser as you move toward the center—eventually denser than liquid water or rock, but still made predominantly out of hydrogen. So defining a "bottom" for Jupiter's atmosphere does not make easy sense. For the purposes of this book, we will consider an atmosphere to be the envelope of gas molecules, ions, and water vapor surrounding a planet (or moon), however tenuous or loosely bound by gravity.

ORIGIN AND EVOLUTION

Why do planets and even some moons have atmospheres? Where do these gasses come from? If planets have atmospheres open to space, a **vacuum**, why don't the gasses simply expand off of the surface and float away? What is responsible for a planet being able to keep an atmosphere, once it has one?

Acquiring an Atmosphere

Planets come by atmospheres in one of two ways: either they managed to capture gasses during their formation from the original **solar nebula**, or they somehow captured or generated gasses later, and built up their atmospheres over time. There are different theories on why the outer planets succeed in retaining large amounts of the original gasses in the nebula, while the inner planets did not. One possibility is that only very large planets had adequate gravity to capture substantial amounts of atmosphere from the solar nebula. Another theory is that the inner planets formed more slowly, and by the time they were large enough to retain gasses, the young Sun entered a very active phase that cleared the gasses out of the interior of the solar system. Large impact events would also have played a substantial role in blasting gasses off newly forming planets. Either way, it is the giant planets of the outer solar system that have retained "primary" atmospheres. The inner, terrestrial planets did not capture these gasses, and so came by their atmospheres through "secondary" processes.

The material that formed into the inner planets included metal and rock as well as more volatile constituents like ices and gasses. Over time, the terrestrial planets differentiated chemically (separated into core, mantle, and crust) and all of them had some form of volcanic activity. These processes released the more volatile substances from the interior of the planets out to their surfaces. This process continues on Earth today, with volcanic activity **outgassing** volatile substances, such as water, carbon dioxide, and sulfurous gasses like hydrogen sulfide into our atmosphere.

Another source of atmospheric gasses is related to impact events. Large impacts early in the formation of the planets certainly caused gasses to be blown into space, but they also might have delivered some volatile materials at the same time, especially if it were a comet responsible for the impact. And for bodies with almost no atmosphere, like the Moon and Mercury, impacts are one of the key processes allowing them to have any kind of atmosphere today. These are tiny impacts for the most part, but they allow small amounts of particles and gas to be ejected from the surfaces of these worlds. On such bodies, the solar wind also plays an important role, with solar wind gasses being implanted and trapped in the upper surface rocks, to be released by later impacts.

On planets with open bodies of water, or with ice or frost, evaporation and **sublimation** can add gasses to the atmosphere. On worlds like Mars, this is not a one-way trip. The gasses sublimate from the ground ice and polar caps in warmer seasons, and then redeposit as frost or polar ice during winter. This means on some worlds the composition of the atmosphere is always changing, back and forth in a cycle, as well as changing more slowly over time with the acquisition of outgassed **volatiles**, etc.

Losing an Atmosphere

Besides an impact event removing atmospheric gasses, or a process like condensation or freezing removing some atmospheric gasses or vapors and emplacing them back onto the surface, a planet can suffer the loss of gasses in other ways, either quickly or slowly over time. Some ways represent a permanent loss, like having an impact blast away atmosphere. Others represent a temporary or potentially temporary loss, like having condensation draw down and freeze water vapor, or a plant uptake CO_2 and incorporate it into its cellular structure. In theory, for each of these last two cases, another process like heating or a chemical reaction could release the gasses into the atmosphere once more.

Technically, other chemical reactions between the planet's surface and the atmosphere are temporary losses. For example, the atmosphere of Mars used to have more oxygen. The oxygen in the atmosphere interacted with the iron in the surface rocks, and the iron oxidized (rusted). This is why Mars appears red. In theory, another process could conceivably liberate this oxygen—but for all intents and purposes, Mars has lost its atmospheric oxygen to its own rocks. The situation is similar on Earth, but altered by the presence of life. Oxygen is also interacting with iron here, and forming rust. But oxygen is also constantly being supplied to the atmosphere by plant life.

The solar wind, while capable of implanting gasses into the Moon's surface, also has enough energy to permanently force some particles from the very top of an atmosphere into space. This particularly affects planets with no magnetic field to shield them from charged particles streaming from the Sun. But the solar wind is not needed in all cases to encourage the loss of gasses from the top of an atmosphere. Planets the size of the inner planets, with their relatively low gravity, cannot hang on to every gas molecule they have. Gas is always, constantly, being lost to space.

The inner planets have enough gravity to hold substantial atmospheres for a long time, but not forever. The larger the planet, the higher the gravity, and the longer a thicker atmosphere can be retained. Not surprisingly, lighter molecules like hydrogen (H_2) are lost more easily than heavier ones like nitrogen (N_2). An element like helium, which is already a very light, noble gas, is lost very easily.

Temperature forms a large part of this picture. Gasses with a high temperature, where the molecules are moving quickly, are more likely to lose particles to escape than those moving slowly. This phenomenon is known as thermal escape, or Jeans Escape. For a single molecule to escape from a planetary atmosphere, it first needs to be near the top, where there are very few other molecules for it to run into, so it can fly freely. Otherwise it is as likely to bounce down from a collision with another molecule as it is to bounce up and out of the atmosphere. Second, the molecule needs to be moving at or above escape velocity. This is the speed required to break free of a given planet's gravitational field:

$$v_{esc} = (2GM/r)^{1/2}$$

This, the standard equation for escape velocity, shows v_{esc} equal to two times the gravitational constant G, and the mass of the planet M, divided by the distance r, all to the ½ power (in other words, the square root).

The velocities of gas molecules go up with temperature, but this does not mean that every single molecule moves at exactly the same speed. Instead, molecules have a distribution of speeds, most of them in the middle of the range, but some much slower, and a few very much faster. So there are always a few molecules at the top of the atmosphere moving at very high speed; one bump in the right direction and they are sailing out of the atmosphere altogether.

The middle of the range is called the "most probable" velocity for a distribution of moving molecules. For a given temperature, this most probable velocity is described as:

$$v_{prob} = (2kT/m)^{1/2}$$

Where v_{prob} is equal to two times Boltzmann's constant k, times the temperature T, divided by the mass m of a single gas particle (molecule or atom), all to the ½ power (the square root). If the v_{prob} is much lower than the escape velocity, then loss of particles will be relatively slow, with only the very fastest in the distribution escaping the atmosphere. But if the v_{prob} is a significant fraction of the escape velocity, that gas will drain away very quickly.

KEY CHARACTERISTICS

Equilibrium

Atmospheres are described by a constant dance, a balancing of opposites. Where, for example, phenomena that **heat** the atmosphere are balanced by phenomena that cool it; where **pressures** that compress the atmosphere are balanced by pressures that expand it, and so forth. The balances are never perfect, as complex forces are always interacting within atmospheres. But the theory of atmospheres is first approached by simplifying, that is, assuming **equilibrium**. Describing equilibrium, and any deviations from equilibrium, then becomes the key to understanding the unique characteristics and changes over time for that atmosphere.

For an example of temperature equilibrium, let's consider an oven. You want to use the oven to bake a cake. The recipe says to preheat the oven. Why bother? Because the cake will bake more evenly if it spends the whole baking time at the proper temperature, and the oven starts at the cool temperature of the room. You turn the oven to 325°F. Is it immediately hot, immediately at 325°F? No. You have to wait five, ten, maybe fifteen minutes or more depending on how good your oven is. You turn the oven to the right temperature, and at that moment the heating elements or coils in the oven grow very hot, increasing the ambient temperature inside the oven.

After a while, the inside of the oven is about 325°F. You put the cake into the oven. Can you now turn the oven off and walk away for 30 minutes and expect a properly baked cake? No. The kitchen is now much colder than the hot oven, and heat is radiating and conducting away from the oven and into the house. If you turn the oven off, it will drop below the proper baking temperature. Instead, you leave the oven on with the dial at 325°F. Now, the coils inside the oven can't heat themselves to 325°F perfectly, in fact, they are much hotter than that. They heat up for a while, then shut off, cool down, and then heat up again. The temperature sensor inside the oven keeps the interior as close to 325°F as it can, based on how it was designed, how fast the coils heat and cool, how well insulated your oven is, and even perhaps if the house is very cold.

This is your oven attempting to maintain a temperature equilibrium. It isn't perfect; as noted; there are heat sources like the coils and also phenomena trying to drain the heat (heat sinks). It takes some time for the oven to turn the coils on and off, and for the temperature to readjust.

But the better your oven is at keeping a steady, nice temperature equilibrium, the more properly your cake will be baked.

For another example, consider pressure. "Imagine a high stack of pillows piled on top of each other. Each has to support the weight of the ones on top of it, so the bottom pillow is squeezed very much, while those higher up are less compressed. Each pillow pushes back with a pressure, and the more it is compressed, the more it pushes back" (Swihart, p. 102). In this scenario, you create a stack of pillows and watch them compress, seeing the bottom pillow flatten more than all the others. And then the compression stops, and you are essentially in equilibrium. The pillows do not have to change thickness anymore because a balance has been reached. And what is doing the compressing? Gravity, of course. If you add more pillows, they will compress more, the one on the bottom getting even thinner. And then, again, the stack will stop compressing as a new equilibrium with the force of gravity is established.

If you don't believe the pillows are pushing back, imagine that you put your hand on top of the stack and add another force. You push down, further compressing the stack. At first it is easy, but the more you push the harder it gets to compress the pillows. Pillows seem light and even flimsy, but they do have internal strength. You push down harder, and the pillows resist, pushing back harder; trying to come to a new balance. You keep pushing down. Eventually you are pushing as hard as you can, and the pillows are still pushing back. You can only push them down so far; you cannot compress the whole stack until your hand is at the level of the floor. Eventually, as you see, the stack will come into equilibrium even with the force of your hand added to the force of gravity.

When a system, like an atmosphere, is in balance between gravity pushing down and pressure pushing back, it is called **hydrostatic equilibrium**. This equilibrium is related to density, temperature, and even the composition of the atmosphere. All these factors work together, pushing back and forth on one another, never in perfect balance, and always in flux. And yet, because of this constant dance around equilibrium, atmospheres behave in consistent and understandable ways.

Pressure

The pillow model is a good model for how atmospheres balance pressure. Instead of a pillow, we have a "parcel" or layer of atmosphere that has to support the weight of the parcels above it. Within each parcel is a number of particles or gas molecules, each constantly moving around and bumping into the "edge" of the parcel. The edge of our imaginary parcel in this case is defined as the molecules in the parcels above and below. Parcels with high pressure are ones where the molecules are bumping into each other and the edges of the parcel frequently, and with more force.

As gravity pushes down on the column of atmosphere parcels, like the stack of pillows, the density (the amount of mass per volume) goes up. This is easy to observe in our model with the pillows, since as the pillows compress you do not see feathers coming out of them. No actual mass has been lost even though the pillow is taking up less space, so the density of each pillow has to be going up. As for our gas parcels, the internal pressure of the gas in each parcel goes up as it is compressed. There is less space, and collisions with the top and bottom edges of the parcel become more frequent. In this way the parcel's pressure rises, and it pushes back against gravity until a balance is reached and the parcel stops collapsing.

As noted, this balance is described as hydrostatic equilibrium. An equation for hydrostatic equilibrium shows the pressure gradient for an atmosphere, in other words, how the pressure changes with distance away from the surface (or center, depending on where you are measuring) of the planet. A simple example for such an equation is as follows:

$$dp/dr = -g\,\rho$$

The dp/dr is a mathematical expression that means "a small change in pressure p over a small change in distance r." The minus sign is there to remind us that the force of gravity is pushing in the downward direction. The g is the symbol for the acceleration of gravity. Rho (ρ) is the density of the parcel of atmosphere. This equation makes sense, even using our simple model of an atmosphere of pillows. How is the pressure of an atmosphere changing over a small distance? It changes as a function of the strength of the gravity field and the density of the atmosphere.

Temperature

With the general idea of equilibrium in place we can begin to look at other aspects of the nature of atmospheres. Having already seen one expression for the relationship between pressure, density, and gravity, we need to include temperature. Temperature turns out to be very important, and one of the most fundamental aspects of understanding atmospheres.

Heat versus Temperature

We often consider temperature to be a measure of relative hot and cold. How cold is it outside? How hot is the oven? And you can look at a thermometer in the first case, and the temperature setting of the oven in the second case, and have some idea of the answer to your question. But heat is not the same as temperature. We equate them in our everyday environment, but in astronomy the real difference between the two is critical.

Both heat and temperature tell you something about the energy of the system in question. But temperature expresses something very specific; it is a measure of the average kinetic energy of the individual molecules in the atmosphere. It describes the amount of motion—generally speaking, high temperature atmospheres have molecules that are moving relatively fast. Heat is not so much a property of a given system as it is a *process.* Heat is the spontaneous flow or transfer of energy between two systems at different temperatures.

When considering temperatures in astronomy, do not imagine temperature as giving you information on how hot or cold an atmosphere might feel. In fact, an atmosphere can have a very high temperature, a high average kinetic energy for the molecules, but still feel cold, even frigid, if you were there yourself. That's because a high-temperature atmosphere, like one near the boundary of space, can still be one with a very low density. Without many molecules to run into you and transfer some energy to you, you will be very cold, indeed. The total heat in such a system might be very small.

The Ideal Gas Law

Using the **ideal gas** law we can incorporate the effects of temperature into our view of pressure and density as follows:

$$p = NkT = \rho RT$$

The variable p is once again pressure, and assuming equilibrium, is equal to N (the number of particles in a given volume) times k (a constant) times T (the temperature) and is also equal to ρ, (the density of the gas, as before), times R (the gas constant appropriate to the type of atmosphere we are talking about) and again, T.

Again, this equation makes sense. If we increase the temperature of a parcel of atmosphere, the pressure must change. The increase in temperature means the average kinetic energy of the molecules in the parcel has gone up; the molecules are moving faster, and running into each other and those in the parcels above and below them more often and with more force. The parcel is pushing outward, that is, the pressure has gone up.

If the weight above the parcel is small enough, then the parcel will expand, the density of the parcel will drop. The molecules will have more room to move, have fewer collisions, and therefore the pressure will be lower. This will establish a new equilibrium with the pressure the same, but with temperature higher and density lower than when we started. If there is a lot of weight above the parcel, it won't be able to expand, and the pressure will stay high. This will establish a new equilibrium with the pressure and temperature higher, but the density the same as when we started.

Scale Height

Our discussion of temperature leads us to the concept of pressure **scale height**. It turns out that the compression of the atmospheric parcels causes the pressure change with height (the pressure gradient) to decrease from the bottom to the top in a very predicable way. For a given planetary atmosphere, the scale height is constant with the temperature.

If you move up in the atmosphere a distance of one scale height, then the pressure will decrease by a factor of 2.72. Go up another scale height and the pressure will decrease by another factor of 2.72. This works for any atmosphere, and so the concept of scale height is a good way to talk about the pressure gradient. Actually the increase is not exactly 2.72, the increase is *e*, which stands for a constant, irrational number, just as the symbol π (pi) stands for a constant, irrational number that begins 3.141 . . . , while *e* is a number that begins 2.718 . . . . Because of this, scale height is often referred to as an *e* folding distance.

Looking at this a little more quantitatively, scale height is defined as follows:

$$H = kT/Mg$$

H is the scale height, and it is equal to the constant k times the temperature T, divided by the molecular mass of the atmosphere in question M, and the gravity at the surface of the planet in question, g. If we consider the atmosphere of a single planet, then g is a constant, as well as M and k. So as noted, H, the scale height, is then a function only of the temperature.

With our previous equations and some math, we come to an expression that makes it more clear why this should be an *e* folding distance:

$$P = P_0 \, e^{(-z/H)}$$

The pressure, P at some height in the atmosphere z, is equal to the pressure at the surface P_0 times, *e*, raised to the power of $(-z/H)$. Looking at Table 5.1, we see an immediate difference between the scale heights of more substantial atmospheres, like Earth, and those very tenuous atmospheres like the Moon and Mercury.

Composition

Now that we have some ways to understand the relationships between pressure, temperature, density, and gravity, we can begin to look at the composition of an atmosphere, and understand how all these variables play together to create the important processes that characterize each planet.

Table 5.1 Key characteristics of terrestrial planet atmospheres, from the thickest to the thinnest.

Planet	Average Surface Pressure	Average Surface Temperature	Scale Height	Composition	Atmospheric Phenomena
Venus	92 bars	878°F (464°C)	15.9 km	CO_2, some N_2	Sulfuric acid clouds, and acid rain in upper atmosphere, very slow winds
Earth	1 bar	59°F (15°C)	8.5 km	N_2 with O_2, some Ar and water vapor	Water clouds, slow to fast winds, water precipitation as liquid and solid ice, storms, dust, other
Mars	6.4×10^{-3} bars Seasonal $(4–8.7) \times 10^{-3}$	−58°F (−60°C)	11.1 km	CO_2, some N_2 and Ar	Thin water and CO_2 clouds, slow to fast winds, storms (no precipitation), dust
Mercury	1×10^{-14} bars	Day/Night 797°F to −283°F (425°C to −175°C)	50 km (Na)*	O_2, Na, H_2, He, and K	No phenomena; gasses captured/sputtered by the solar wind
Moon	1×10^{-15} bars (night)	Day/Night 257°F to −283°F (125°C to −175°C)	1022 km (H)* 120 km (Na)* 90 km (K)*	He, Ne, H, Ar, and K	No phenomena; gasses captured/sputtered by the solar wind

**Note*: Scale heights for Mercury and the Moon are not easily comparable to the other planets. The rarified atmospheres mean that each component must be dealt with separately, and measurements for some of these constituents have large error.

Air versus Atmosphere

When discussing atmospheric composition, it is important to point out the difference between what scientists mean by "air" and what they mean by other, related terms like "atmosphere" or "gasses." Air means something very specific to a scientist; it is only found on Earth, and is the combination of breathable gasses and water vapor that make up the bulk of our atmosphere, i.e., mostly nitrogen with some oxygen, and then a little argon and water vapor. This is generally what a scientist means by the word "air."

Martian Rock, Martian Gasses

Knowledge of the Martian atmosphere helped solve one of the great mysteries in meteorite science. One meteorite group, known as the SNC meteorites, showed evidence of being from an object much larger than any asteroid (where most meteorites arise) and an object that experienced volcanism much more recently than any asteroid. There were reasons to suspect these meteorites came from Mars, but scientists could not figure out how Martian rocks could make the journey to Earth intact, and concluded they must be from a different object.

In the 1980s, trapped gasses were found in volcanic glasses in SNC meteorites. These gasses were found to be an excellent match for the gasses in the Martian atmosphere, known from *Viking* lander data. With this match, it was conclusively shown that the SNC meteorites originated on Mars, and within a decade other scientists determined a way for them to have arrived. Study of these meteorites, now known as "Mars meteorites," has provided great insights into Martian history and conditions as a result of knowing where they formed.

Air is not the same as oxygen. Oxygen is one possible gas, either as an ion (O) or a molecule (O_2) that is found in planetary atmospheres. Wind is also not air. Winds can arise on any planet with a significant atmosphere, and are driven by changes in pressure and temperature. There are winds of nitrogen and oxygen on Earth, winds of CO_2 on Mars, and winds of hydrogen on Saturn.

The Greenhouse Effect

The phrase **greenhouse effect** is often misunderstood. It is sometimes equated with the phrase, **global warming** and thought to be both a human-created phenomenon, and one that also negatively impacts the planet. But the greenhouse effect is a natural and important function of our atmosphere, and responsible for our planet being habitable.

The basic idea is straightforward. In a sense, the molecules in our atmosphere act as a blanket, helping to trap heat and keep it closer to the planet longer than it would remain otherwise. Visible light from the sun passes through our atmosphere almost unimpeded. It strikes the ground, where some is immediately reflected away, and some is absorbed. This absorbed visible light is reradiated as infrared (IR) light. Unlike with visible light, the molecules in our atmosphere are more likely to absorb IR radiation, and some molecules are particularly good candidates for this. The gasses that easily absorb IR radiation are called greenhouse gasses and include carbon dioxide (CO_2), water vapor (H_2O), and methane (CH_4). Once absorbed, the molecule once again will emit the IR radiation, but of course it has the same opportunity to emit the light down as up, toward space, where another molecule will absorb the IR and continue the process. And so the greenhouse gasses ensure this IR energy

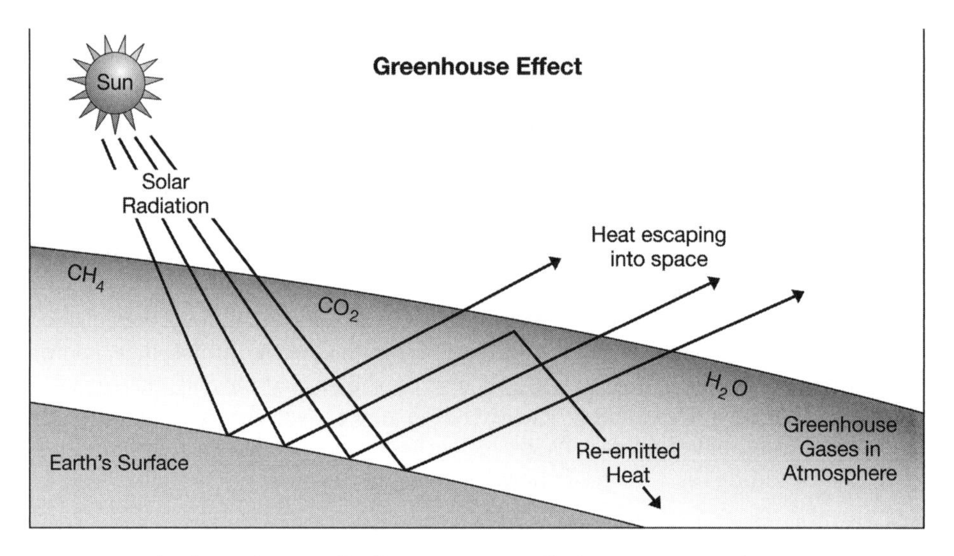

Figure 5.2 The "greenhouse effect" is present on all planets to some degree. At present it warms the Earth and allows comfortable temperatures over much of its surface. Current theories suggest that the addition of carbon dioxide to the atmosphere over the last century or two will increase the Earth's temperature due to an enhanced greenhouse effect, potentially leading to global warming and climate change.

remains in the atmosphere for an extended period, keeping the surface of the planet warm.

Note the surface temperatures listed in Table 5.1. For the first three planets, those with substantial atmospheres, only an average temperature is listed because, to some extent, they all have a greenhouse effect caused by their atmospheres. This means that even after the sun sets, the temperature of the nighttime half of the planet does not drop substantially. It may seem counter-intuitive since we have all experienced it getting colder at night. But looking down the table makes it clear that such an observation is relative. Mercury and the Moon have almost no atmosphere, and certainly no greenhouse effect. Day and night temperatures are listed for these bodies, and illustrate how dramatic the temperature shifts can be without an atmosphere. On Mercury, the difference is more than 1,000°F from high daytime to low nighttime temperatures! The greenhouse effect helps to smooth out the wild swing of temperature variations to something tolerable for life.

It is possible, however, for a planet to have a greenhouse effect that is so pronounced it warms the planet past the point life can exist. Venus is such a case. It has a very thick, dense CO_2 atmosphere, very effective as a greenhouse gas. Almost all the IR emitted by the surface is trapped by the atmosphere. Temperatures on the surface exceed 450°F, easily high enough to melt lead. And there is no difference at all between the day and night sides—the heat cannot escape. The middle of the night is as hot as high noon.

You might wonder at Mars's lower surface temperature, after all, it has CO_2 as the major constituent for its atmosphere, as well. But Mars is also further from the sun and has a much thinner atmosphere. It is enough to help buffer the day and nighttime temperatures, and keep the planet above the frigid nighttime temperatures of −175°F we see for the Moon and Mercury. But it is not enough to drive the temperatures even as high as the Earth's.

WINDS AND ATMOSPHERIC CURRENTS

Weather and Climate

Like "air" and "atmosphere," the words "weather" and "climate" are often used incorrectly or interchangeably. The basic difference between the two is one of time. Weather, as we described at the beginning of this chapter, is something that happens in the short term, tomorrow or perhaps next week. Weather is essentially the day-to-day changing of temperature, precipitation, pressure, wind speed, and more, that make up the picture of our near-surface atmospheric environment. Climate is how these short-term changes average out over a longer span of time. For example, it does not rain often in the hot desert Southwest of the United States. The climate is

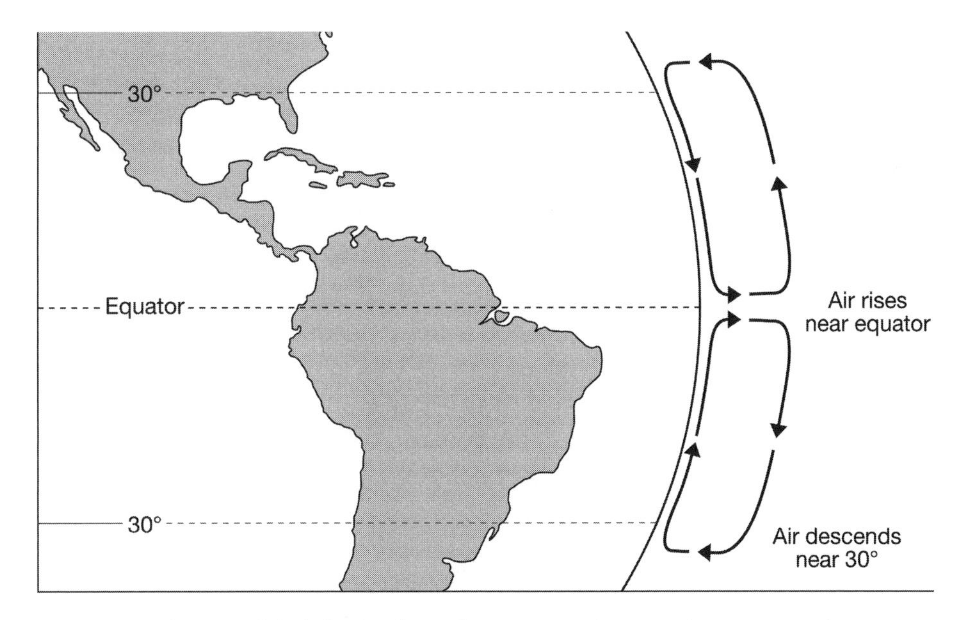

Figure 5.3 This simplified figure shows how convection acts in an atmosphere. Hot parcels of air near the surface at the equator become less dense and rise, pushing other air out of the way. Once they have risen, they give off their heat and start to cool. Once they have cooled, they begin to descend, now further from the equator after having been pushed away in their own turn. Finally, back near the surface they move equatorward to take the place of air that has risen, heating up as they go and beginning another cycle. These cycles, shown in cross-section as circles, are called "Hadley Cells." Similar cells exist further away from the equator as well.

hot and dry, year after year. But the weather does vary; you can get rained on in the desert. Nevertheless, you will experience rain less often, with a total amount of precipitation much less than you would have in, say, the colder and wetter climate found in the New England states.

Sometimes trying to find the balance between what is weather and what is climate can be tricky, and this is part of the reason that global warming remains a controversial topic. It is not immediately obvious if a change in weather patterns is temporary, or if it actually represents the start of a long-term, critical change into an era with a very different climate. A temporary change in weather patterns can be difficult, such as a drought for several years running that jeopardizes farming in a region. But of much more concern is if such a drought is not happenstance, but instead representative of a new climate, where the region in question has become permanently unsuitable for farming.

Global Air Circulation

Atmospheres would be far less interesting than they are without a powerful energy source driving the changes within them. Weather, for example, is only possible because the atmospheres of the terrestrial planets are energized by the Sun. However, the Sun is not solely responsible for all the details of the weather we see. Air currents and such are the product of heating and cooling combined with the effects of a planet's rotation, seasonal variations, and much more.

The first effect to consider is that of solar heating. Imagine the Earth. We know what will happen to a parcel of air as the temperature goes up—the pressure within the parcel rises along with it. The density of the hotter air parcel goes down, and it becomes lighter than the surrounding air. This is a more detailed way of stating the axiomatic "hot air rises." As warmer air from the equator moves upward, it flows north and south to the polar regions, cooling as it goes. The air parcels lose altitude as they increase in density, descending at the poles. At this point, the air has no place else to go but back down and around to the equator, being too cool to rise and also being pushed by more air coming in behind it. This is the simplest model of an atmospheric circulation cell.

Of course the Earth is rotating, and this affects these large-scale circulation cells. The Earth's rotation, via the **Coriolis force**, causes the one large air circulation cell to break into a few smaller **Hadley Cells**. How does this happen?

The air near the equator is moving faster than the air at the poles because it is farther away from the axis of rotation for the Earth. Imagine a spinning top. If you could shrink down and stand upon it, you would find you would be whipping around much faster standing near the edge of the top than near the center. The result for the Earth is that the air at the equator moves out with a slightly higher average velocity than the rotation rate of the Earth, while air moving in, toward the equator, is moving slightly slower. The difference causes the air flowing in the southern hemisphere to tend clockwise, and the air flowing in the northern hemisphere to tend counter-clockwise. This is the Coriolis force, and is responsible for the opposite rotation direction of hurricanes in the southern and northern hemispheres.

The result for our circulation cells is that air cannot move directly north-south to and from equator to pole and back again. The air in each hemisphere moves with westerly winds at high latitude, with easterly winds at mid-range latitude, and westerly again near the equator. The net effect is the creation of smaller Hadley convection cells driving the large-scale air currents across the Earth. Because of these large-scale, global wind and air circulation patterns, we can predict the direction weather systems will move around the globe.

CONCLUSION

Our common understanding of atmospheres derives from our everyday experience with phenomena like the weather. However, the atmospheres of other planets are diverse, and no one explanation for their origin and evolution can be applied to all of them. In spite of this, fundamental principles and characteristics are common to all atmospheres, such as equilibrium, and the interactions between pressure, temperature, composition, and density. As a result of heating from the Sun and the effects of planetary rotation, winds and weather unique to each planet arise and enact changes on the surface of the world.

FOR MORE INFORMATION

Bennett, Jeffrey, Megan Donahue, Nicholas Schneider, and Mark Voit.*The Cosmic Perspective.* 4th ed. San Francisco: Pearson Education Inc., Addison Wesley, 2007.

Chamberlain, Joseph W., and Donald M. Hunten. *Theory of Planetary Atmospheres: An Introduction to Their Physics and Chemistry.* 2nd ed. Burlington, MA: Academic Press, 1987.

Swihart, Thomas L. *Quantitative Astronomy.* Upper Saddle River, NJ: Prentice Hall Inc., 1992.

6

Rocks as Clocks: Radiometric Processes

INTRODUCTION

People want to know what time it is. If asked "What is the function of a clock?" most people would answer, "To tell the time." The advent of digital, even voice-responding clocks makes this seem obvious. However, while the reason you buy a clock is to know what time it is, the function of the clock is in fact to accurately measure time passing. For example, an analog clock must be designed such that it takes exactly one second for the second hand to move one tick around the clock face. Imagine you have two clocks, both reading 7:15. Each ticks 60 times, and now one reads 7:16, and another reads 7:55. One of your clocks is not accurate. You can not rely on it to tell the time because a second to the clock is not the same as a second for you.

Good clocks are **calibrated** to ensure that it takes one second for the second hand to move just far enough to measure out a true second. So a clock is not actually telling time, it is measuring, as accurately as it can, the amount of time that has passed. A clock is designed to be very, very predictable, or it has no use. When you first buy a clock you do not necessarily expect that it has been set properly. You check it against another timepiece to be sure it is telling the right time. If not, you change it to match. If you don't do this, you may look at your new clock during the day and see it reads 10:00 p.m. You know that is wrong, but for the moment you can not know exactly what time it is because you don't have another clock to reference. But, you can still look at your new clock later and note that it now reads 11:00 p.m. So you know that whatever time it is, an hour has passed since you last looked. The clock is not malfunctioning even though it is not telling the correct time; it is working well and predictably. The clock just needs to be matched to a known **standard** before it can give you the exact information you want.

A stopwatch is a more simplified version of timekeeper. Such a watch is not used to give the time of day at all, but to find out how much time passes between when you hit "start" and when you hit "stop." Then you can hit "reset" and do it all over again. Once you reset the clock, it has no memory of the previous time interval. It is back to 00:00, ready to go again.

You can use a stopwatch to tell time. Again, what you need is a standard. In this case, you choose to reference a friend's watch. Exactly at noon, you hit start, and the stopwatch goes. From that point, all you or anyone needs to know is when it started (noon) and a little arithmetic and you can calculate the time. If it reads 49:15:06. You know that two days, one hour, 15 minutes and six seconds have gone by. So it is about 1:15 pm, two days after you started. A stopwatch is clearly a clock, measuring the passing of time in a nicely predictable fashion.

As it turns out, rocks can be clocks, too, in a way much like stopwatches. Rocks contain certain elements that spontaneously, over time, turn into other elements. These are the **radioactive** elements, and they **decay** into other elements naturally, as a matter of course. The original, individual radioactive atoms are called **parents** and they decay into **daughter atoms** or **daughter products**. Like the reset and start on a stopwatch, if you know how many parents you started with, and either how many parents are left, or how many daughters have been produced, you can calculate how much time has passed since this geologic clock was "reset." Perhaps you can determine the formation time of the rock itself. Radioactive decay happens very predictably, and so scientists can use this phenomenon to learn a lot about rocks, and about the geologic contexts in which they are found.

RADIOACTIVE DECAY

Some elements are naturally radioactive, that is, they are not **stable** over some duration of time that varies by element. Radioactive elements are found naturally in all aspects of our environment, and they are decaying around us all the time. They can be found in animals, plants, rocks, rivers, the atmosphere and inside of people, just to name a few. While it is possible to artificially increase amount of radioactive elements in a rock or person, for example, you cannot eliminate all radioactive elements from the environment. They are a normal part of nature.

Radioactive decay is rarely as simple as one **unstable** parent atom decaying to one stable daughter atom with the emission of some kind

Determining the Age of the Earth

While the rate at which elements decayed has been known for some time, knowing the initial mix of parent isotopes was a much harder problem to solve. Without that information, scientists could measure the age of rocks but could not determine the age of the Earth itself without making several assumptions. By the 1930s, initial attempts at dating the Earth via measurements of uranium and helium led to an age of at least 2.5 billion years. This was later increased to 3 billion years.

It was not until the 1950s that the problem was finally solved. Studies of the Canyon Diablo meteorite using a variety of lead isotopes pinned the age of that meteorite at 4.55 billion years old, and it was argued that that was also the age of the Earth. Studies over the subsequent 50 years have supported and refined that age.

of energy. A very large atom might split into many pieces, or only decay in stages. Atoms might decay into daughters that are themselves radioactive, and so on down the line, until eventually a stable daughter is reached. In the process, the decaying atoms will release energy at various wavelengths, as well as particles. The energy and particles emitted are commonly referred to as "radiation," although it is very important to note that scientists use this word to mean many different specific phenomena.

What makes this natural radioactive decay so useful is that the process is very predictable. Strangely, no one can tell when any *one single atom* will decay. If you isolate a radioactive atom you will find you cannot predict when it will decay, it happens randomly. Yet, if you have a large population of these atoms, *the group of them* will decay in a highly characteristic fashion.

Scientists use the term **half-life** to express how fast a population of atoms decays. One half-life is the time it takes half of the original parent atoms to decay. Some radioactive elements have very short half-lives, hours or less, while others have half-lives of many billions of years. The data for the half-lives of all radioactive elements can be read from a **chart of the nuclides**. This chart is, in a sense, the full expansion of the periodic table of the elements, showing all isotopes of all elements. (Data here are from the *Interactive Chart of the Nuclides* produced by the National Nuclear Data Center.)

We begin a hypothetical experiment with a million atoms of Cesium-137 (^{137}Cs). ^{137}Cs has a half-life of about 30 years. Ideally, after one half-life of ^{137}Cs, or 30 years, we would have 500,000 parent atoms of ^{137}Cs left. After another half-life, 250,000 are left, and so on. This data is shown graphed in Figure 6.1.

The bulk of the ^{137}Cs is gone in three half-lives, but a small amount continues to persist for a very, very long time. That is the nature of all decay curves. If we add other elements (from Table 6.1), and expand the "time" axis much further, we get the data graphed in Figure 6.2.

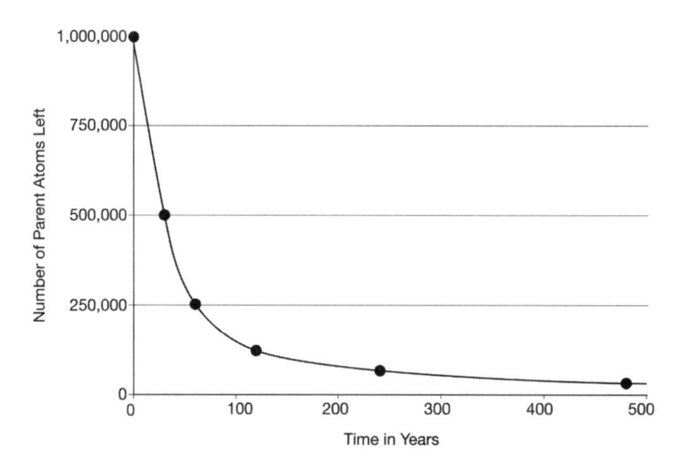

Figure 6.1 Ideal exponential decay curve for Cesium-137 with a hypothetical data set of Cesium-137 atoms starting at 1 million.

In 10,000 years (more than nine half-lives) the ^{137}Cs is essentially gone. But on the other extreme at 10,000 years, Plutonium 239 (^{239}Pu) has yet to experience a single half-life. More than a tenth of our starting sample of ^{239}Pu will still be around after 100,000 years.

These curves are "exponential" curves, and can be described by the following equation:

$$N/N_0 = e^{(-\lambda t)}$$

N is the number of parents that remain after some time *t*, and N_0 is the number of parents we started with. The symbol *e* stands for a constant, irrational number, just as the symbol π (pi) stands for a constant, irrational number. π is a number that begins 3.141 . . . while "*e*" is a number that begins 2.718 . . . The letters *ln* stand for **natural log**, which is the same as the standard function *log* but to the base *e* instead of base 10. λ is the decay constant, which relates to the half life ($t_{1/2}$) as $t_{1/2} = ln(2)/\lambda$.

So . . .

$$(t_{1/2}/\ln(2)) \times \ln(N/N_0) = -t$$

Table 6.1 Half-life of selected radionuclides.

Isotope	Half-Life in Years
Cesium-137	30
Mercury-194	450
Carbon-14	5,700
Plutonium-239	24,000

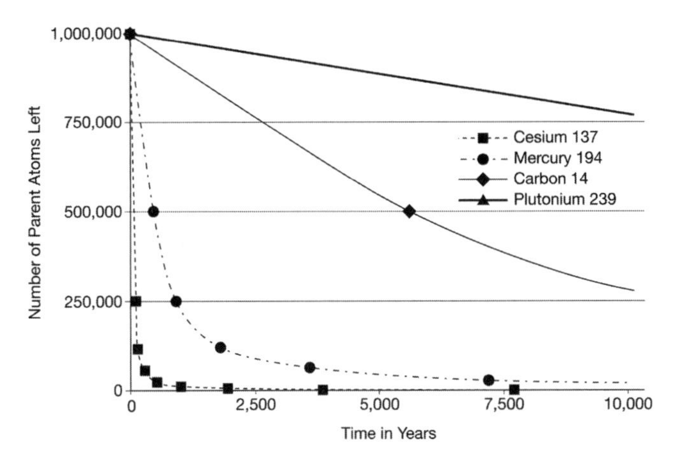

Figure 6.2 Exponential decay curves for Cesium-137, Mercury-194, Carbon-14, and Plutonium-239 with hypothetical data sets of parent atoms starting at 1 million.

Taking the carbon-14 system as an example, if we start with 100 carbon-14 atoms, and wait until only 25 are left, we know that two half lives have gone by, or over 11,400 years. To confirm we have the following:

$$(5730/\ln(2)) \times \ln(25/100) = -t$$
$$8267 \times -1.386 = -t$$
$$11458 = t$$

(For a full derivation of equations for radioactive decay, and a more rigorous treatment of this subject see Faure's *Principles of Isotope Geology* [1986].)

CARBON-14 DATING

The most well-recognized dating system is probably the Carbon-14 system. This particular system is very useful for dating organic matter that is no older than about 50,000 years or so. As you can see from the above graph, by 50,000 years, there would be almost no Carbon-14 (^{14}C) left in the object being tested. This is more than eight half lives for ^{14}C. Measuring any difference in age beyond that point would be very difficult or impossible because of the extremely high precision required for the necessary measurements. It is also important to note that this system cannot be used to find a very young age, either, since it would be very hard to measure the small change in ^{14}C over that amount of time.

The Carbon-14 system has been used in many famous capacities, such as dating the Shroud of Turin and the Dead Sea Scrolls. The system is used widely in physical and cultural anthropology to date all sorts of fossils and relics, including shells, bones, wood, fabric, even eggshells. Anything that once was a part of a living creature and drew carbon in from the environment as it lived is a possible candidate for Carbon-14 dating.

Note that Carbon-14 cannot tell you how old a tree was when it died, although you might determine that from other means. The Carbon-14 "stopwatch" starts when the tree dies. This system, like a good clock, can tell you *how long it has been* since the tree died. It is very important to understand what it is that starts and resets radiometric stopwatches. What is it that we are actually measuring, so that we know how much time has passed? As mentioned previously, we need to know three factors to calculate an age; the first is how many parent atoms we started with, the second is how many parents we have left (or we can count how many daughters have been produced, and subtract to find how many parents are left), and third we need to know the half-life of the parent population of radioactive elements. With this information, we can calculate how much time has gone by since the start. But what is the start time? Why does the Carbon-14 clock start when the tree dies?

Carbon is readily available in Earth's atmosphere, and it comes in various isotopes, as do all elements. Most of the carbon is stable Carbon 12 (^{12}C), but a very small amount is radioactive ^{14}C. The ratio of ^{12}C to ^{14}C in nature is essentially constant (normally, for every one ^{14}C atom, there are roughly 1,350,000,000,000 or $1.35x10^{12}$ ^{12}C atoms). Therefore, if you know how much ^{12}C is in a sample of air, you can use this ratio to calculate how much ^{14}C is there, and vice versa.

All carbon has the opportunity to interact with oxygen in the atmosphere and become carbon dioxide (CO_2). Plants, algae and the like absorb CO_2 and incorporate it into themselves as a part of the process of photosynthesis. When other creatures eat plants and are then eaten by other animals, the carbon ingested now becomes a part of the eater. As long as a tree or animal is living, it takes in carbon. Some of this carbon is ^{14}C, which is decaying as this process goes on. But since the tree is alive, it continually recycles carbon within itself. Therefore the ratio of ^{12}C to ^{14}C in the tree remains constant; the same ratio as it is in the atmosphere.

This is akin to a stopwatch that is constantly being reset. Some of the ^{14}C in the system begins to decay, and the watch starts. But immediately, the reset button is pushed, as the living organism takes in more carbon, and resets the ratio of ^{12}C to ^{14}C back to the original ratio. For a living tree, the time measured by the radiometric stopwatch is zero with regards to ^{14}C. Scientists refer to this as an "open" system, one that allows the isotope of interest (in this case ^{14}C) to generally remain free with regards to the sample that is of interest (in this case, the tree). In an open system, either no decay can be measured because the parent isotope continues to be replenished; or the daughter isotope is not being contained so that it can be counted, etc. As we know, we have to count either the parent or daughter, know the half-life, and know the starting ratio of isotopes to determine how much time has passed.

For the tree, its Carbon-14 system "closes" when it dies. At that moment, the stopwatch starts and does not get reset again. As the ^{14}C

continues to decay, it no longer gets replenished. The ratio of ^{12}C to ^{14}C in the tree starts to change. The more time passes, the more ^{14}C is lost, and the higher the ratio goes. At any time, a scientist might take a sample from the tree and measure the amount of ^{12}C to ^{14}C; and knowing the half-life of ^{14}C, the scientist can calculate the time since the tree died. These ages are approximate, as are all radiometric ages, because the systems are complex. For example, the ratio of ^{14}C to ^{12}C in the Earth's atmosphere has actually varied a little over time; it has not been perfectly constant.

To use the Carbon-14 method of dating, a scientist takes a sample from a dead tree, or other organic matter of interest. Several different techniques and tools exist for the scientist to process the sample and learn the carbon content, including **mass spectrometry**, **Geiger counters**, **scintillation** techniques, etc. The end result is that the scientist measures the ^{12}C and ^{14}C present in the sample. In this example, the scientist finds 5,400,000,000,000,000 or 5.4×10^{15} ^{12}C atoms. Because the ratio in nature is constant, this means the number of ^{14}C atoms in the beginning (N_0) is:

$$\text{Known ratio } N_0/5.4\times10^{15} = 1/1.35\times10^{12}$$
$$\text{so } N_0 = 4000$$

From this same sample, the scientist measures the number of ^{14}C atoms remaining (N) as 500. How long has it been since the tree died?

$$(t_{1/2}/\ln(2)) \times \ln(N/N_0) = -t$$
$$(5730/\ln(2)) \times \ln(500/4000) = -t$$
$$8267 \times -2.079 = -t$$
$$17191 = t$$

Is this age consistent with everything we know? Looking at the starting and ending amounts of ^{14}C, we know that three half-lives have gone by (since 4,000 to 2,000 is one half life, 2,000 to 1,000 is a second half life, and 1,000 to 500 is a third). So we expect an age of:

$$3 \times 5730 = 17190$$

Therefore our estimate of 17,191 years is consistent; the tree died about 17,190 years ago.

FINDING THE AGES OF ROCKS

Since the basics of the Carbon-14 system are straightforward, it provides a good starting point for understanding more complex systems, such as those used for finding the ages of rocks. In spite of the increased complexity of these systems, certain steps and principles remain the same, no matter what system is being used. The scientist conducting a study must:

1. choose a dating system appropriate both to the rock sample they have and the questions they are trying to answer,
2. understand what "event" is being dated within the history of the rock (that is, what has "reset" the radiometric stopwatch in the rock),
3. understand what assumptions have been made in the dating system being used, and how their specific case may or may not deviate from the ideal case, and,
4. correctly interpret and apply the ages they have determined to the geologic context from which they got their rock sample, or they will not be able to answer the questions of interest to them, or not answer them correctly.

An Example Using the Potassium-Argon (K-Ar) System

Imagine scientists are interested in studying old, cooled lava flows on an island much like Hawaii. These types of lavas generally erupt in a very fluid form of molten rock, which cools rapidly after it reaches the surface. These scientists are therefore interested in knowing the last time the rocks of the cooled lava flows were molten.

The first step, choosing an appropriate dating system, is based both on what elements are in the rock, and what system might tell you when a rock was last molten. Volcanic rocks such as those in Hawaii contain plenty of Iron (Fe) and Magnesium (Mg), as well as some other elements like Potassium (K) and Calcium (Ca).

Is there a system that tells you when a rock was last molten (melted)? Yes, the **Potassium-Argon system**, K-Ar, can be reset by melting, so it can tell you the last time a rock was in a molten state. Matching the fact that the rocks in question contain enough Potassium (K) to measure, and that the K-Ar system detects the event of interest to the scientists, this radiometric system makes sense for the study.

Measuring Isotopes

The instrument most commonly used to measure the isotopic mix in a sample is a mass spectrometer. These work by taking advantage of the differing masses of isotopes. First, the sample is ionized, vaporizing its atoms and giving them a net electric charge. An electromagnetic field inside the instrument accelerates the ions, which move at different speeds due to their differing masses. The separation of the ions can then be measured either by noting the different times they arrive at a fixed location or by observing them move differently in the electromagnetic field. Armed with this information, the relative amounts of ions of different masses can be calculated, and from that the isotopic ratios present in the sample.

Table 6.2 K-Ar age data for a hypothetical set of samples from an imagined island. Six lava flows were sampled, with individual rocks labeled with letters. The table lists the estimated age of each sample and the analytical error.

Lava Flow and Sample	Age (years)	Analytical Error (range of probable age)
1 A	1.41×10^6	+/– $.07 \times 10^6$ (or 1.34 to 1.48 $x10^6$)
1 B	1.17×10^6	+/– $.09 \times 10^6$ (or 1.08 to 1.26 $\times 10^6$)
2 A	1.32×10^6	+/– $.01 \times 10^6$ (or 1.31 to 1.33 $\times 10^6$)
2 B	1.37×10^6	+/– $.05 \times 10^6$ (or 1.32 to 1.42 $x10^6$)
3 A	1.19×10^6	+/– $.08 \times 10^6$ (or 1.11 to 1.27 $x10^6$)
3 B	1.10×10^6	+/– $.02 \times 10^6$ (or 1.08 to 1.12 $x10^6$)
3 C	1.16×10^6	+/– $.05 \times 10^6$ (or 1.11 to 1.21 $x10^6$)
4 A	0.68×10^6	+/– $.03 \times 10^6$ (or 0.65 to 0.71 $x10^6$)
4 B	0.50×10^6	+/– $.03 \times 10^6$ (or 0.47 to 0.53 $x10^6$)
5 A	0.56×10^6	+/– $.06 \times 10^6$ (or 0.50 to 0.62 $x10^6$)
6 A	0.49×10^6	+/– $.02 \times 10^6$ (or 0.47 to 0.51 $x10^6$)

The second step, related to the first, is to understand what it is in the history of the rock sample that is really being dated. The scientists in question wish to explore when the rocks erupted as liquid lava, and then cooled. But how can a radiometric system be sensitive to when a rock was molten?

Potassium-40 (^{40}K) decays into two different, stable daughter products, Argon-40 (^{40}Ar) and Calcium-40 (^{40}Ca). Argon is a noble gas, and once created it will not interact with other elements or minerals. If K40 exists within a rock, over time, the amount of K40 will go down, and the amount of ^{40}Ar will go up. But the noble gas argon is not "bound" into the minerals of the rock. In a sense it is formed in place when a ^{40}K atom decays, and then is stuck there. When the rock melts, the argon formed within the rock becomes liberated, and rapidly leaves (**degasses**). This degassing, this release of the argon, resets the K-Ar clock. When the rock cools, the clock "starts," with more ^{40}K decaying into argon, and more argon building up in the rock until such time as the rock is melted again.

Using our stopwatch analogy, as long as the rock is molten, no ^{40}Ar builds up. It is as if we keep hitting reset on the watch over and over before any time can register as passing. Once the rock has cooled sufficiently, the ^{40}Ar begins to get stuck again, and builds up in the rock. The radiometric stopwatch has begun to tick in its very precise fashion.

Radiometric systems allow us to *estimate* ages with varying degrees of confidence. All radiometric systems are based on assumptions, or are modeled around an ideal case. Depending on the accuracy of our measurements and how much the rock we want to analyze has deviated from the assumptions we have made, or deviated from the ideal case, we may calculate a very accurate age estimate, or not.

In our imagined study of island lava flows, the scientists visited the island, mapped out the lava flows, and took rock samples from each flow. They found six lava flows, and took anywhere from one to three samples from each flow. Using the K-Ar technique, they found the ages of these rock samples, and compiled Table 6.2.

The scientists made several immediate observations from their table. The two ages for Flow 2 overlap within error from 1.32 to 1.33 million years. The three ages for Flow 3 overlap within error from 1.11 to 1.12 million years. Only one sample could be obtained for both Flows 5 and 6, so the best estimate for those ages remain 0.5 to 0.62 million years for Flow 5 and 0.47 to 0.51 million years for Flow 6.

Flows 1 and 4, however, are not internally consistent. The ages for the two samples for each of these flows are not even close to overlapping. What has happened?

At this point the scientists take into consideration the third principle from above, which suggests that not all experiments will be ideal. Something has happened to change the experimental situation for Flows 1 and 4 so that they no longer approximate an ideal case. One or more of the assumptions upon which the K-Ar system is built has been violated.

Understanding what has happened to these samples brings us to the fourth point: the need to interpret the age data properly in light of the geologic context. The scientists refer to Figure 6.3, their map of the lava flows on the island:

Examination of the map shows that samples 1B and 4B were collected very close to other lava flows. In each case, a younger flow (3 and 6) erupted and flowed over an older flow (1 and 4) respectively. It appears that when the later hot lava flows passed over the old, cold flows, the heat allowed some or all of the argon in the older rocks to degas. This effectively changed the apparent ages of those rocks to something much younger than it should be. To be consistent with the overlap on the map, the overall geologic context, and the K-Ar data, the scientists take the ages for 1A and 4A as the most likely ages for those flows, not including 1B and 4B in this determination.

Using all the data to put together a story, the scientists ultimately learn the following about the volcanic history of the island. About 1.41 million years ago, a large flow erupted from the edge of the summit caldera of the volcano. It was followed by another large flow dated at 1.32 million years, and a third large flow dated at about 1.11 million years. Three large flows, all erupting from the summit of the volcano over a period of 0.3 million years. The next flow, Flow 4, does not erupt until 0.68 million years ago, more than 0.5 million years after Flow 3, and it did not erupt from the near

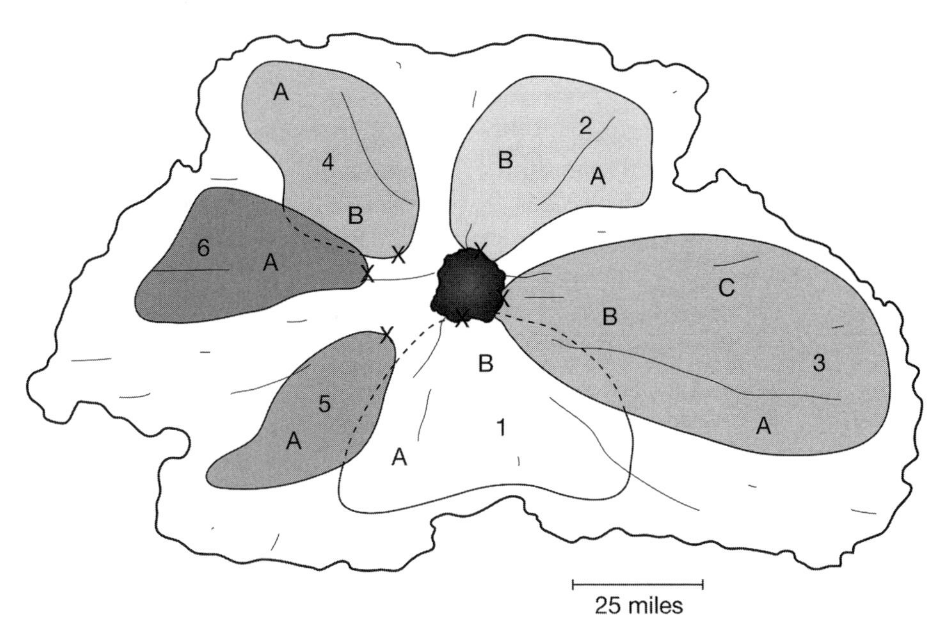

Figure 6.3 This is a map of a small hypothetical Hawaiian-type island. The center circular feature is the top caldera of a volcano. Six lava flows have been mapped on the surface of the island, apparently flowing downhill to the sea. The numbers indicate the different flows. The x symbols show the points where lava flows originated. The letters show the locations where rock samples were collected.

central caldera, but instead broke out from a crack in the side of the volcano. Flows 5 and 6 followed Flow 4 relatively quickly, at 0.56 and 0.49 million years, respectively. Like Flow 4, these were somewhat smaller flows, and also erupted from the side of the volcano. Three smaller flows, all erupting from the west, northwest flanks of the volcano over about a 0.2 million year period. It seems that volcanism on the island happened in two major episodes, with much larger flows in the earlier episode. A fundamental shift happened before the second episode of eruption, since these flows are much smaller and originate from the flank of the volcano and not near the summit.

This example shows the usefulness of rocks as clocks. Without the age data, the volcanic history of the island could not be understood in this detail. The rocks have measured the passing of time, like very precise stopwatches. As long as we understand how to read these clocks, what causes them to "start" and "reset," we can use them to help unlock the geologic story of their surroundings.

Other Geologic Radiometric Dating Systems

There are a host of systems in routine use in geologic investigations. Each one offers unique aspects, and so scientists are careful to choose the system best suited to the rock samples they have and the problems they are interested in investigating.

The K-Ar and Ar-Ar (Argon-40/Argon-39) systems are both based on the decay of ^{40}K to ^{40}Ar, although in the Ar-Ar system, ^{39}Ar stands in as a proxy for the parent ^{40}K. In each case, it is the retention or release of argon that controls the system. Rocks which have K in their minerals, and which have been melted and cooled completely are very good candidates for these techniques. However, these systems can also date rocks that were not completely melted. For example, after a rock is formed, it may be covered with **sediments**, **metamorphosed** (heated) below the surface, and then **uplifted** in a mountain-forming event. In this case, the heating while the rock was metamorphosed may have allowed some of the argon to escape from the rock. The Ar-Ar technique can be used in some of these cases to date the time the rock was **exhumed** (the mountain-building event) rather than the original formation age of the rock. The Ar-Ar technique is widely used in planetary science to date thermal events such as impacts or impact crater formation on planetary bodies, or within meteorites.

The U, Th-Pb (Uranium, Thorium-Lead) systems are very complex, as uranium experiences a sequence of decays, producing radioactive daughters including thorium, before ending with stable lead daughter products. These systems are Uranium-235/Lead-207, Uranium-238/Lead-206, and Thorium-232/Lead-208. Minerals with adequate quantities of these elements, and which retain all the various daughter products over time can be analyzed to determine when they went from being a melted part of a liquid reservoir (perhaps below the surface) to a discrete body of crystallized rock. Zircon is a mineral commonly analyzed using these systems. The crystallized mineral "locks in" the amount of U and Th within it when it crystallizes. Note that this does differ from the cooling ages estimated by the K-Ar and Ar-Ar methods, not only in that different minerals are appropriate for the different systems, but also because a rock can be crystallized, closed to U and its daughter products, while still being hot enough that it cannot retain Ar, that is, still "open" to Ar.

The Rubidium-87/Strontium-87 (Rb-Sr) system is also useful for calculating the crystallization ages of some rocks. An Rb atom is about the same *size* as a K atom, so Rb can be found in the place of K in K-bearing minerals like feldspar. Assuming that a rock sample does not undergo a thermal event that mobilizes or *disturbs* the Rb or Sr within it, and assuming an appropriate choice of an initial Sr ratio, this system can be used to estimate the crystallization age for the rock. Measuring and understanding the changes in Sr ratios over time allows for the investigation of a suite of phenomena, including the nature of meteorite parent bodies, the evolution of the Earth's mantle, and changes in ocean water composition.

The Samarium-147/Neodymium-143 system (Sm-Nd) can occasionally be used to date rock samples that are not suitable for Rb-Sr, perhaps because those rocks have undergone some event that has kept them open in the Rb-Sr system. Model ages within the Sm-Nd system estimate the time that has gone by since the sample in question had the same Nd composition as some of the most primitive material in the solar system. Using Sm-Nd, scientists can model the evolution of material over time, as its Sm-Nd differs more and more from its initial value. Similar is the Lutetium-176/Hafnium-176 (Lu-Hf) system, in that it also is sensitive to change from the initial ratios of these elements from the primitive material found in meteorites.

CONCLUSION

Rocks are the clocks that can tell us how long it has been since a wide variety of different geologic events have taken place, both on Earth and in space. The clocks inside of rocks are based on the natural and highly predictable process of radioactive decay of certain elements. The clocks within rock samples operate like stopwatches, measuring the time since the watch was "reset." Each radiometric dating system is reset in a different way, and is appropriate for different rock samples depending on composition, and the question under investigation. Radiometric clocks in rocks can estimate the ages of a variety of events, including: the time when a lava flow cooled, the age of a mountain range, the time an impact crater formed, the crystallization age of meteorites, and more. Data from rock-clocks, used in concert with other data available to scientists, such as **stratigraphic** data, **seismic** data, **geochemical** data, and detailed maps, allow scientists to understand the stories behind the major events in solar system history.

FOR MORE INFORMATION

Hawaii famously has very active volcanoes, of the sort used as examples for age dating in this chapter. The Hawaii Volcano Observatory (http://hvo.wr.usgs.gov/) maintains a set of web pages discussing past and current activity as well as webcams.

Information extracted from the Chart of Nuclides database, http://www.nndc.bnl.gov/chart (hosted by the National Nuclear Data Center) was used in this chapter, a site that also includes a glossary to aid understanding the chart and its use.

Newman, William. Geologic Time, available online at http://pubs.usgs.gov/gip/geotime/contents.html, is a classic consideration of the age of Earth and materials on it. (Published by the United States Geologic Survey [USGS] Publications Warehouse, http://infotrek.er.usgs.gov/pubs/, which also includes numerous other publications from the USGS.)

7

Triple Point: Hydrologic Processes

INTRODUCTION

Humans are enraptured with water. It forms the basis of our obsession with weather, our love of hot baths, and our enjoyment of walks along the shore. We drink it, sled on it, swim in it, fish from it, and wash with it. And we fear it in the form of rainstorms, icy roads, and flooding. Water in all its phases of vapor, liquid, and ice is one of the most central substances around which our lives revolve. Life as we know it cannot exist without water.

Water has been and continues to be a critical substance in shaping the surfaces of the inner planets. While only the Earth still has a substantial quantity of water, enough to create a true hydrosphere, the other worlds have all been affected by the presence or lack of water during and after their formation. Water changes rocks in important and predictable ways. It also flows over the surface of planets both as a liquid and as icy **glaciers**, altering and eroding the land beneath.

Liquid water is the most easily recognized form of water for most people on Earth. There are climates where ice rarely forms, if ever. Climates that are both hot and humid, such as jungles, have abundant liquid water, as well as abundant water vapor. Although you cannot see it, you can certainly feel it in the form of humidity. Water vapor is invisible—clouds and steam are not actually water vapor, but very tiny droplets of liquid water suspended in the atmosphere. Solid water is easily recognizable in colder climates as snow, ice, sleet, and even accumulating over time into glaciers. Glaciers are solid ice, and yet they do flow under the right pressures and temperatures. And solid water possesses a very strange property; it is less dense when in solid form than when liquid. Almost all other substances are denser as solids, but this is not the case with water, which is why ice floats.

Important questions remain about how water is delivered or accreted to planets during formation, how it changes the nature of the planetary formation process, and what causes water to be lost from planetary environments. Why is it that the Earth has such a great quantity of this fantastic substance, while planets both closer to and further from the Sun have so much less, or none at all? Why is it so important that we live on a planet near the triple point of water, and what does that mean for the formation, evolution, and persistence of life?

THE TRIPLE POINT

We know that water (H_2O) comes in three simple forms: solid (ice), liquid (water), and gas (water vapor). We happen to live on a planet where all three phases can exist at the same time. If Earth were closer to the Sun our water may have boiled away to be nothing other than a trace of vapor in our atmosphere. Further from the Sun, and our water may have frozen solid, with no liquid left over to allow for life. Earth is a planet near the **triple point** of water, and by this we mean that the natural conditions of temperature and pressure on the Earth's surface fluctuate around an important point for this substance.

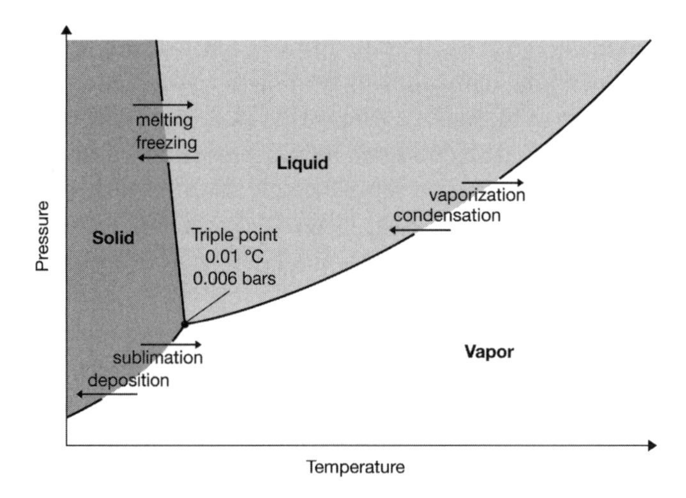

Figure 7.1 The physical state of water changes with pressure and temperature. Between 0 and 100°C at 1 bar of pressure water is a liquid, for instance. Physicists and chemists have observed and calculated the state of water for a variety of pressures and temperatures. Remarkably, the Earth's surface can have conditions that allow all three states of water to exist simultaneously, called the triple point.

The phase diagram shows at what pressure and temperature you can expect to find each phase of water. In our everyday environment, we are familiar with the change that occurs when temperatures drop below 32°F, or 0°C: liquid water **freezes**. If the temperature is then raised above this value, the ice **melts**. We are also familiar with the way water beads up on the outside of a cold glass. This happens when water vapor in our atmosphere **condenses** on the cold glass to form water droplets. It is commonplace to see **evaporation** in action, with puddles or spilled water "drying up" over time. Frost forming on a cold window is an example of **deposition**, and when frost disappears directly into the atmosphere without first melting, that is **sublimation** in action.

Because of this wide variety of phenomena, water has a vast number of ways to interact with planetary surfaces, both mechanically and chemically. *Mechanical* in this context refers to physical processes such as abrading, cracking, and pushing, which are all part of how water erodes landscapes. *Erosion* is a dynamic physical process that differs from *chemical* forms of water interaction such as weathering and aqueous alteration of minerals. Given that Earth's climate allows water in all of its phases, the Earth shows the full range of water's chemical and mechanical changes to the landscape. But the other inner planets are by no means devoid of the effects of water on their surfaces or in their atmospheres. Mars had vast quantities of water on its surface in the past and possesses ancient dried river channels as evidence. And Mars still retains a small amount of water trapped as ice in its soil and in the caps at its poles. Venus's atmospheric chemistry suggests it had plenty of water in the past, but lost it through processes that contributed to the creation of that planet's thick, hot atmosphere.

CHANGING LANDSCAPES

As we know from living on Earth, water changes the environment. The most straightforward way it does this is by the mechanical processes of **erosion**. As water flows over the surface of a planet, it works, either slowly or sometimes catastrophically, to wear down existing structures, eliminating them or perhaps creating new ones in the process. Glaciers can also flow, if more slowly, and ice can break rocks apart as water freezes inside cracks and pore spaces. Water is a highly effective sculptor of landscapes.

Rivers, Lakes, and Oceans

If a planet has rain, or has a dramatic melt of ice either from the near surface or from the polar regions, then liquid water can flow in sufficient abundance to cause channels, rivers, lakes, and even oceans. In the case of the Earth, rain and ice melt are common. Because of this, water has ample

Figure 7.2 A photo taken of the Grand Canyon in Arizona, one of the most famous water-carved features on the Earth's surface. Over millions of years the Colorado River eroded the landscape resulting in the canyon we see today. As opposed to the Colorado River, it is thought the water on Mars spent a very short time above ground, perhaps as ice that melted quickly due to proximity to a volcano and burst forth from underground.

opportunity to arrive at the surface by one means or another and flow downhill. The act of flowing slowly cuts through the rocks and soil, first forming rivulets, then streams, then rivers as the water seeks the topographic low point. As the water moves, it picks up bits of rock and other debris. This makes the water even more efficient at erosion since entrained sediment acts as a scouring agent. Fast flowing, turbulent water can entrain a great deal of sediment, and is very effective at cutting its way across planetary surfaces.

As water moves over and carries along rocks and sand, it works to smooth them. You may have noted that pebbles near streams have a rounded or "tumbled" look. Rocks that have been mechanically weathered in such a fashion provide some evidence for water flow, either present or in the past. Entrained sediment is carried efficiently by fast-moving water, but as rivers widen and the flow rate slows, less sediment can be carried.

Figure 7.3 A combination of compositional and topographic data have identified these craters in the Hellas region of Mars as containing glaciers, covered with Mars dust. The ice, currently not visible through the dust, spills out of the bottom crater and downhill.

The river "drops" some or all of its carried sediment, and this can form new features, or even fill in the bottoms of rivers and lakes over time.

Eventually, the water reaches the local topographic low point. It is retained there, forming a lake, sea, or even an ocean. Erosion does not end here, since waves and tides continue to move the water against the shoreline, creating distinctive features. Oceans can be long-lived, as they are on the Earth. The pressure and temperatures on Earth are conducive to stable liquid water at the surface. Mars may well have had oceans and lakes in the past, but that water is now gone. Temperatures and pressures on Mars are too low to allow liquid water to persist.

No water system acts completely independently of other factors. Hydrologic processes work together with atmospheric, volcanic, tectonic, and other processes to produce dramatic water action. Rain, rivers, and oceans can appear to be slow, consistent landscapers. But they do most of their work during sudden and relatively short episodes. Atmospheric storms can be accompanied by large amounts of precipitation in a short period of time, which can cause rivers to swell and flood. Faster-moving water entrains more and larger particulate matter, and on Earth this can include vegetation. Such swollen rivers have a devastating ability to erode the landscape. Ocean cutting is also more efficient during storm events as waves batter the shorelines. On Mars, episodes of volcanic activity may have caused large amounts of subsurface ice to melt catastrophically. The massive Martian outflow channels could be a result of several episodes of heating and melting. **Tsunami** are waves created in oceans by major tectonic events such as earthquakes and landslides. These towering waves can instantly reshape a coastline, and on Earth cause widespread damage.

Glaciers and Ice

Liquid water is not the only efficient landscaper. Ice can dramatically alter the surface of a planet. On smaller scales, ice can form inside the pore space of rocks or in cracks, and as it repeatedly freezes and then melts, it slowly acts to pry rocks apart and to widen cracks. **Frost heaves** are places in the ground where subsurface ice has forced the soil upward. This small-scale effect slowly accumulates over time, and eventually can be responsible for dramatic landslides and other phenomena.

On larger scales, ice can accumulate into wide expanses known as glaciers. Glaciers can appear to be solid and unmoving, but they are actually quite mobile and under the right conditions can flow easily, if slowly, across a planetary surface. Glaciers can entrain a great deal of large-sized particulate matter. Because of this they can be highly effective in mechanical erosion, wearing down sharp, pointy mountain peaks into rounded hills. In addition to this erosion, glaciers create a wide variety of distinctive land features as they drag rocks, sand, and dust along with them and deposit them on the landscape. Some features have been identified on Mars that are consistent with glaciated terrain on Earth, indicating the possibility of localized glaciers on the Red Planet.

The polar regions of any inner planet are a place to look for possible water ice (and perhaps other volatiles in their solid form). Planetary poles

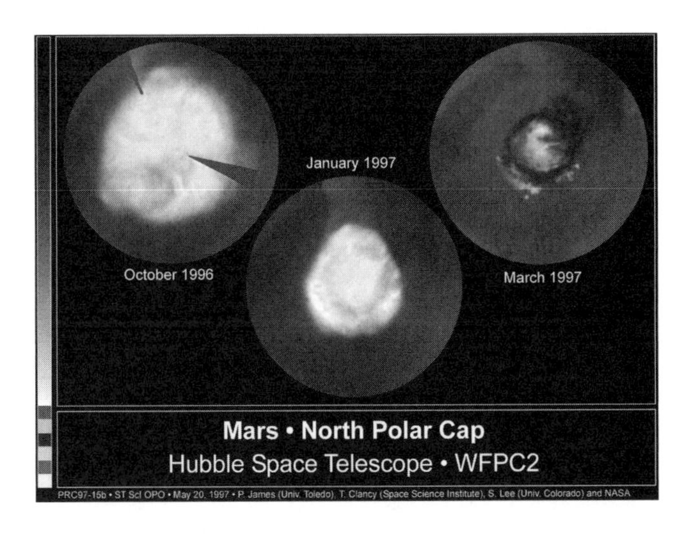

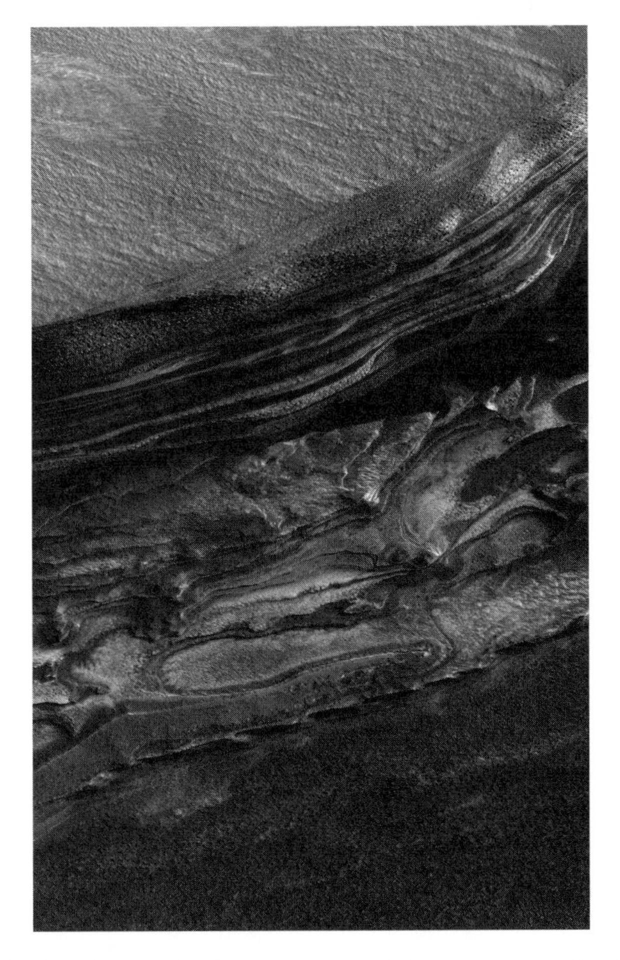

Figures 7.4 Mars' polar cap has been seen to shrink and grow with the seasons for hundreds of years. The top panel shows Hubble Space Telescope images of the north pole of Mars as its cap grows and shrinks. The constant advance and recession of the cap causes a unique set of layers and terrains, seen on the bottom in an image of the cap's edge from the *Mars Reconnaissance Orbiter*.

receive the lowest levels of direct sunlight, and so can become reservoirs for water frozen out of the atmosphere or sequestered in permanently shadowed craters. The Earth and Mars both possess substantial ice caps, and in the case of Mars both water and carbon dioxide are frozen in quantity in the polar regions. The Martian polar caps grow and shrink dramatically with the Martian seasons, creating unusual layered formations.

ALTERING ROCK CHEMISTRY

In addition to the more obvious kinds of mechanical erosion, water has an important chemical effect on rocks. Water can act to remove, or **leach** minerals from rocks. It can become bound into the structure of a rock or interstitially between rock layers, creating entirely different minerals. It can deposit minerals out of solution and thereby form brand new rocks. And it can act along with heat to create the unique products of hydrothermal alteration.

Water as a Solvent for Rocks

Water is a very effective **solvent**, so effective it is often referred to as the "universal solvent," although that can be misleading since certainly water cannot dissolve everything. Solvents are substances that serve to break apart the chemical bonds in other materials. The process of creating a **solution** is more involved than simply breaking a material into smaller pieces and then mixing those pieces with water. The dust from a crushed rock floating in a glass of water is not **dissolved** into the water, it is merely dust suspended in liquid. If you had a filter with small enough holes you could strain out the rock and have exactly what you started with—bits of rock and a cup of water. When a solvent (like water) acts to dissolve a **solute** (like sugar) it forms a solution, which is a new kind of substance altogether. A solution cannot be separated back into its starting ingredients by simple mechanical means. The new substance has properties that differ from the old. For example, when salt is dissolved in water, the freezing temperature of the water goes down. This is why salt is used to inhibit the freezing of snowy roadways in winter.

Solvents act electrochemically to break the bonds within the molecules of some other substance. That substance, the solute, could be a solid, another liquid, or even a gas. Common solvents are most often liquid, but gasses, and even in extreme cases some solids, can act as solvents. What matters are the electrochemical properties of the bonds between the molecules of the solute, and if and how the solvent that is being used is effective in breaking those specific bonds.

For example, water is highly effective at dissolving ionic compounds like salt. An ionic compound is essentially a group of ions or molecules bound together by a positive and negative charge. One part of the ionic compound is electromagnetically charged positive, while the other is charged negative. They therefore each attract and maintain themselves as a discrete—separate—material. When such a material, like sugar for example, is placed in an appropriate solvent, like water, the solvent acts to break the bond. Water can do this to sugar because water is a polar molecule. A water molecule is like a little magnet, since the oxygen lends a net negative charge to one end of the molecule, and the two hydrogen atoms lend a net positive charge to the other end of the molecule. Sugar has hydroxyl groups (OH) that are somewhat negatively charged, forming the negative part of the ionic compound. When a small amount of sugar is placed in water, the sugar is surrounded by highly ordered (aligned) water molecules. The charges holding together the ionic compound are outnumbered and overwhelmed by the strong polarity of the water molecules. Instead of being attracted only to the positive group in the ionic compound, the hydroxyl groups are also attracted (and perhaps even more strongly attracted) to the positive ends of the many water molecules. This encourages the sugar to break apart and dissolve into the water. After the sugar breaks apart, it does not have a tendency to reconnect, since the negative groups of the former compound are surrounded by positive water charges, and the positive groups of the former ionic compound are surrounded by negative water charges, effectively stabilizing the two different groups. Even very strong ionic bonds in some compounds and molecules can be broken by the overwhelming action of water as a solvent.

Dissolving and Precipitating Minerals

We have seen above how substances like salt and sugar dissolve when poured into a cup of water. These same types of processes occur in nature. Instead of a salt shaker and a cup of water, we can imagine a floorless sandbox filled with a mixture of sand and salt. Over time, rainfall into the sandbox will dissolve a small amount of the salt and carry it along out the bottom of the sandbox and into the ground, but leave the sand alone. After a long enough period, measurements of the sand to salt ratio in the box will show a very different result than what was originally put in.

This same process can occur in rocks. Water can seep in via preexisting cracks or pores in the rocks themselves. The combination of water and some minerals, like carbonates, can create a weak acid that can dissolve other minerals. Small cracks can become larger cracks through this process, and voids can form. Indeed, many of the largest caves on Earth formed via this type of process.

The other half of this process is deposition. Water carrying dissolved material may change temperature or salinity, or may evaporate entirely.

In such a case, the dissolved material may be left behind. Returning to caves, the stalactites and stalagmites that are often found in limestone caves are consequences of deposition, as carbonate-rich water drops its dissolved minerals on contact with air, slowly building up large masses with time. Some features found by the *Mars Exploration Rovers* and nicknamed "blueberries" have been interpreted as due to the deposition of minerals once dissolved in water, though this is not yet settled.

Water Within Rocks

Water itself can be incorporated into minerals. The electromagnetic properties of water discussed above serve to provide a thin layer of **adsorbed** water molecules on most Earth materials. This water can be dislodged by heating or long-term exposure to very low pressures and very low humidity. Some minerals have crystal structures that make it easy for them to incorporate relatively large amounts of water. For instance, anhydrite ($CaSO_4$) is transformed into gypsum ($CaSO_4 \cdot 2H_2O$) upon reaction with water. Montmorillonite, a commonly found mineral on Earth, can incorporate large amounts of water, with an increase in volume occurring with the water intake. Some of the most common minerals on Earth and Mars, including montmorillonite, are **weathering products** formed by the interaction of water and volcanic minerals. The water in these weathering products is often incorporated between layers of other minerals. This *bound water* is more difficult to remove than adsorbed water, requiring much higher temperatures.

In addition to those minerals that incorporate water are some minerals that just incorporate the hydroxyl ion (OH), derived from water. These minerals, which can also incorporate water, also tend to be weathering products. Hydroxylated minerals usually have very strong bonds with their

Water, Water Everywhere?

Human beings use a lot of water. We drink it, use it to cook, and use it for hygiene. It is also used for growing our food and cooling our equipment, among many other uses. Unfortunately, it can be expensive to ship water into space for use by space travelers, and a great deal of effort has been expended to try and minimize requirements for water, as well as recycle water that might normally be discarded.

The interpretation of ice near the lunar poles has led planners to suggest future lunar missions to be sited there, to allow astronauts relatively easy access to water. The incorporation of water into minerals has led some scientists and engineers to propose sending future astronauts to Mars with tools to extract bound water for their use, more difficult work than simply melting ice/dust mixtures and filtering the water. The process of extracting water from hydrated minerals, called in-situ resource utilization (ISRU), has been studied with increased interest. The possibility of hydrated minerals on asteroids and the Martian moons has also been proposed as an opportunity for ISRU.

OH, and require temperatures of hundreds of degrees Celsius to lose the OH. This often will destroy the mineral entirely.

Hydrated minerals (those with water) and hydroxylated minerals are ubiquitous on Earth and Mars. They have been searched for on the Moon, with promising results as of this writing. These searches are usually based on remote sensing in the infrared spectral region, where water has a very strong absorption.

Water and Heat

The chemical reactions involved in weathering, like most chemical reactions, become more efficient at higher temperatures. The temperature and pressure conditions found at planetary surfaces can be very different from those found in interiors, which can potentially change the results of chemical reactions as well as their rate.

Rocks are altered by high-temperature water in **hydrothermal systems.** These systems tend to be associated with volcanic centers, whether large mid-ocean ridges interacting with ocean water or smaller magma chambers heating groundwater. Because the specific minerals created in hydrothermal systems and their relative amounts are dependent upon temperatures and pressures, identification of minerals can help scientists determine and trace the history of a rock.

WATER AND LIFE

You have probably heard the statement that life as we know it requires liquid water to exist. This is the reason that searches for life in the universe center around a search for water. But why is liquid water so important?

The Composition of Life

Life as we understand it is very complex. No one definition exists that describes it or allows us to identify it without ambiguity. We know from basic biology that most forms of life have certain characteristics such as: growth, reproduction, metabolism (the use of fuel), adaptation (change in response to the external environment), and homeostasis (regulation of a stable internal environment), among others. Life is generally self-organizing and self-producing. But as always, there are exceptions and very simple forms of "life" that fall on the boundaries.

But certainly every kind of biological life requires chemical reactions to exist. Chemical reactions form the basis of how an organism structures itself, keeps itself separate from the outside environment, uses fuel, promotes change within itself, and allows for growth and reproduction. Life is therefore suspected to be more likely in those environments rich in the chemicals that allow for sophisticated reactions and the liquid solvents that allow for those chemicals to dissolve, mobilize, and then recombine. And naturally there will be particular temperatures and pressures conducive to those reactions, as well as appropriate energy sources of use to the life form.

Life on Earth is "carbon based," that is, carbon provides the essential link necessary to combine other abundant elements like hydrogen, nitrogen, and oxygen in highly complex ways. Carbon can do this because it is has particular physical properties that gives carbon four unpaired electrons ready to bond with other elements. Because of the number and strength of the bonds, carbon can be used to create very long and complicated chains of molecules. These chains allow for all the highly complex biochemistry necessary for life as we understand it on Earth. Indeed, carbon is so important that an entire large branch of chemistry, organic chemistry, is centered on study of that element.

This does not mean that life cannot theoretically be formed around elements other than carbon. It does mean that there are no other elements that can fill carbon's particular niche quite so well. It isn't impossible that a related element, like silicon, could perform a similar function. Silicon sits below carbon on the periodic chart, and therefore also has four unpaired electrons available to combine with other abundant elements. But silicon is larger, and does not form certain bonds as easily as carbon. Silicon simply does not interact with other elements in exactly the same way as carbon—especially when it comes to interactions with oxygen. Oxygen is abundant in the universe and highly reactive, which means it can be quite toxic to life. If you are a life form and oxygen gas is on your planet, then you need to adapt or evolve clever ways of using it or protecting yourself from it. Higher forms of life on Earth take in oxygen and create carbon dioxide as a waste product. Carbon dioxide is a gas, and so it is easy for a life form to eliminate it from its cells. But the analogous process for a silicon-based life form would likely involve the creation of a solid, something much more difficult to eliminate.

So it is not that carbon is the only possibility, but it appears to be the most favorable candidate for the basis of life for a huge host of reasons. And there is quite a lot of circumstantial evidence in favor of carbon-based life as opposed to silicon (or some other element). For example, prebiotic molecules and chains of carbon are actually quite commonplace in the solar system. Many kinds of complex chains of organic molecules have been found in meteorites and comets. While some silicon molecule chains are out there, they are few, and many of those also incorporate carbon. Another piece of circumstantial evidence is that in spite of silicon being far more abundant than carbon on the Earth, it is carbon that forms the basis of life as we know it. It would appear that the creation of the initial building blocks of life, as long as it is carbon-based, is

The Origin of Earth's Water?

Water is one of the most common substances in the solar system. The satellites of the giant planets are covered with it, and it dominates their masses. Scientists studying the formation of the solar system, however, are still uncertain whether its presence on the Earth should be a surprise. Ice is not stable on airless bodies at 1 AU, the Earth's distance from the Sun. While it is stable on objects with atmospheres, like the Earth, it is hard to figure out how to have formed the Earth without beginning from small airless bodies on which ice was unstable.

One of the possible answers to this problem comes from the small bodies of the solar system: asteroids and comets. Meteorites, which largely come from asteroids, can have up to 10–20 percent water by weight incorporated in their minerals. Comets are known to be icy bodies. A substantial amount of water could have been delivered to the Earth late in its formation by these impacts. Exact details are still being calculated by modelers, who are trying to figure out whether this scenario can fit the data.

straightforward. (Of course, what turns these chains from simple molecules into part of a complex living organism is something scientists still do not understand.)

With these and many other issues to consider, it seems prudent for scientists to concentrate on understanding and searching for carbon-based life in particular. This directly leads to searching for liquid water, because water is an excellent solvent for carbon molecules.

Water as a Solvent for Life

As with carbon versus silicon, water is not the only option imaginable as a solvent for life. Ammonia also possesses certain properties that could allow it to provide a similar function. But it does appear that water is the most versatile solvent, with unique properties that can support the functions of life in a variety of conditions.

But why does life need a solvent? Life as we know it operates inside of small units called cells. Each individual cell is a highly complex chemical factory, metabolizing energy and molecules, creating waste products that must be efficiently eliminated, and generating other useful materials for the body, and then getting those useful materials off to the places they need to be. All of the chemical reactions inside this factory are related to one another, with one reaction usually spawning another, and then another, in a chain that ends up with the appropriate product for the cell. Because of this, all of the molecules, elements, and compounds need to be mobile enough to move around and interact both with each other and other parts of the cell, but not so mobile that they fly apart and never bump into one another. A cell therefore needs an effective solvent to encourage and support these generally carbon-based chemical reactions. A liquid is called for, since a gas would not constrain the interacting molecules well enough, and a solid would not allow enough freedom of movement. And as it happens, most of the organic molecules associated with life are nicely water-soluble.

Water performs other services within a cell, as well. Large carbon molecule chains can be encouraged to grow when dehydrated, and then broken apart when rehydrated. And it is because water cannot effectively dissolve lipids (fatty soluble molecules) that allows the lipids to separate water into neat cellular packets. One end of certain lipids is water-friendly (hydrophilic) while the other end dislikes water (hydrophobic). These lipids therefore line up to form membranes where the hydrophilic side neatly surrounds and traps water, and the other side faces other hydrophobic lipids. The chemistry of life has much to do with how different products can be encouraged to enter and leave the cell at appropriate times while maintaining the hydration environment within.

All of this leads scientists to believe that, first, life probably needs an effective solvent, and this solvent needs to be able to perform many functions. If the life is carbon-based, then water is the only liquid that does all of these things for carbon molecules that is also found at the temperatures and pressures needed to power life.

As noted, it is not impossible for other solvents to sustain life. If life were acting at a lower temperature, then perhaps other polar solvents, such as ammonia, would be effective. But ammonia does not support the same sort of metabolism found in life on Earth. It is possible that ammonia and water together could form a sort of "antifreeze" in organic life, a means for it to survive and operate at lower temperatures. But alone it does not provide the most encouraging option for a solvent for carbon-based life. Even sulfuric acid could possibly sustain some kind of metabolic reaction for carbon life. However, a strong sulfuric acid is a good solvent for other materials, like rock itself, which can cause other problems for life.

A Last Word About Life

As noted, it is by no means impossible for life to be based on an element other than carbon, utilizing a solvent other than water. Some of the more bizarre life on Earth pushes these boundaries. It is a very big universe, and the speculations of science fiction do have their place as a means to use our imaginations to expand our horizons. But science fiction is not *science.* Science requires that results and conclusions be based in observations and evidence, that they can be predicted and repeated, and that they stand up to repeated tests of their veracity. This is why the search for life is so closely tied to the search for water, and does not try to look systematically for all possible kinds of life on all possible worlds. Such a search would be impossible anyway, since we have no hope of searching through all the worlds in the galaxy. Scientists must find a way to narrow down the candidates to make any kind of methodical search plausible. We might not even be able to identify life if it is very different from ours, so searching for it—when there are other, much more robust and verifiable options available—is not the best use of resources. Within our own solar system, we have the luxury of sending actual probes and perhaps stumbling upon very unusual forms of life. But for systematic telescopic searches of worlds far beyond our own, we need to be selective. There is a very strong theoretical basis for the presence of water on a planet, in the past or present, as a link to the possibility of life. We know without a doubt that carbon/water life can exist on an Earth-like planet, so it is not simple lack of imagination or an overly "Earth-centric" point of view that leads us to look for similar worlds. It is taking the most robust and authentic observations and theories that we have about life, and applying them in a way that gives us some hope of finding it elsewhere someday.

CONCLUSION

Water is an amazing substance. It is a relatively simple molecule and commonplace in the universe, and yet it acts dynamically on the inner planets to alter landscapes through mechanical erosion and alter rocks and minerals through aqueous chemical processes. Water is found in all three states within the inner solar system, especially on Earth, where vapor, liquid water, and ice have all played a role in evolving the planet's surface. It seems that liquid water is pivotal in allowing for the development and persistence of life. Scientists continue to search for water within our own solar system, as well as in planetary systems around other stars, in the hopes of understanding how planets form, and perhaps to find the hallmarks of life.

FOR MORE INFORMATION

http://www.lsbu.ac.uk/water/index2.html: Technical discussion of a wide range of scientific considerations of water.

http://vulcan.wr.usgs.gov/Glossary/Glaciers/framework.html: United States Geological Survey collection of glacier-related web links.

http://ga.water.usgs.gov/edu/index.html: Basic information about water.

8

Planetary Shield: Magnetospheric Processes

INTRODUCTION

At the center of our solar system is a brilliant star. Even at a distance of 93 million miles, the energy output of our star, the Sun, is so vast that it warms the entire inner solar system. This star is responsible for providing the energy that makes life on Earth possible.

But there is a flip side to having such a powerful energy source available to us. The Sun is dynamic, constantly changing, sometimes unpredictable, and capable of creating flares and even storms of radiation and charged particles that blast out into the solar system and slam into the planets. In addition to more violent events like solar flares and coronal mass ejections, there is a wave of charged particles flowing off of the outer atmosphere of the Sun at all times. This wave is called the **solar wind**, and it interacts with everything in its path. Under the circumstances, it is amazing life has flourished on our world.

One of the reasons that life has done so well here is that we have some protection from the Sun. Our atmosphere, for example, helps to block much of the ultraviolet radiation that would otherwise pass unimpeded right to the surface. But our atmosphere cannot completely stop a blast of charged particles. Fortunately for us, there is an invisible shield surrounding our planet that can do just that, at least in part. It sounds like something from a comic book—an invisible barrier protecting life on Earth from dangerous solar activity.

This shield is generated by our own planetary magnetic field. You've no doubt encountered bar magnets, and seen how they can force small iron filings around them into a pattern of curved lines. These lines are caused by, and trace out, the magnetic field created by the magnet. A planet can also be a magnet, creating a vast, invisible field to surround it, many times bigger than the planet itself. Charged particles flowing from the Sun encounter this magnetic field, and instead of plowing forward into the Earth, the majority of them become entrained in the Earth's magnetic field and are forced around and away from the planet. Because of this, and many other factors, the presence or absence of magnetic fields around a planet can have profound effects on its formation and evolution, and on the ability of life to thrive there.

RELATING ELECTRICITY AND MAGNETISM

It is very difficult to understand magnetospheres—how they are generated and what their effects are—without understanding the basic nature of electricity and magnetism. In our everyday experience, the two phenomena we call **electricity** and **magnetism** may not seem related. Most people encounter magnets or magnetic materials as part of toys or items that stick to the refrigerator. Electricity is commonly understood to be "what runs the appliances and lights," often without considering how that is actually accomplished. Therefore it can be surprising, and very nonintuitive, to find out that electricity and magnetism are two sides to the same coin. There are both **electric fields** and **magnetic fields**, where a **field** in this case can be considered to be the region or sphere of influence of the force in question. (Technically, a field extends indefinitely throughout all of space, but its strength can decrease quickly with distance.) These fields can be considered separately to start, but are so closely intertwined that they cannot be truly understood until they are considered together.

Electric Fields

Any **electrically charged particle**, by its nature, is surrounded by an electric field. As we know, a proton and an electron are naturally electrically charged particles; the proton is positively charged and the electron negatively charged. If two electrons enter one another's electric fields, a force will be exerted on the particles. In the case of similar (like) charges, the force will act to repel the two of them from one another.

This is the basis of how electric fields are useful to us in our everyday lives. Imagine we have a good **conducting** material, like a metal wire. A natural property of metal is that it has plenty of free electrons, which are not bound to specific atoms, but are instead roaming about within the

North and South

The Chinese invented the compass roughly a thousand years ago, while Europeans created the modern compass in the 1300s. While the Chinese convention was to have a south-pointing needle and the Europeans a north-pointing one, the practice of naming the ends of a magnet "north" and "south" was an obvious path to take, with the north pole of a magnet being attracted to the Earth's north pole.

The first proposal that the Earth itself was a magnet occurred in the 1600s, and as magnets became better understood in later years, it was recognized that because like poles repel, the north poles of magnets must be attracted to the *south* magnetic pole of the Earth! By this point, however, the names were too well-ingrained, and it is now accepted that the north magnetic pole of the Earth is actually a south magnetic pole. Some people use + and – instead of "north" and "south" to try and reduce any confusion.

metal matrix. When this wire is under the influence of an electric field (perhaps by having each end connected to a battery) the electrons in the wire are subjected to a force and immediately move in one direction. In the case of a battery, the electrons will be forced toward the positive end. The movement of electrons is responsible for the **electric current** that powers all of our electric devices and makes modern life possible.

Magnetic Fields and Magnetic Materials

In much the same way that an electron naturally has an electrical charge, it also has a **magnetic moment**. This is an intrinsic property of an electron arising from its motion—both its spin, and its tiny orbit around the nucleus of an atom. An electron moving in this fashion will generate a magnetic **dipole**, that is, it will become its own individual mini-magnet. Any magnet, even an electron, has a magnetic field surrounding it that bends from one end, called the north **magnetic pole** of the magnet, to the other end, the south magnetic pole.

Diamagnetism

Of course we know that most items in our common environment are not magnets, or at least are not magnetic enough to stick to the refrigerator. But in a sense, everything is magnetic, because everything has electrons and atoms that will, if only weakly, respond to a magnetic field. In most materials, the angular moments from spin and from orbit cancel out, so all that remains is a very weak form of magnetic behavior called diamagnetism. **Diamagnetic** materials will respond by magnetically opposing any outside magnetic field and by being repelled by either pole of a magnet.

However, this reaction will be extremely small, much too small to be seen without exceedingly sensitive instruments and exceedingly strong magnets. Once the magnetic field is removed, the material will generally return to the way it was before. Examples: wood, plastic, plants, water, and even metals like gold.

Paramagnetism

A smaller subset of materials are **paramagnetic**. In these materials, the application of an outside magnetic field will cause electron spins to align, causing the material to be attracted to a magnet. Paramagnetism can range from very weak to moderate in strength, with the strength of the response in proportion to that field. Examples: atoms of the elements oxygen, sodium, tungsten, and platinum.

Ferromagnetism

An even smaller subgroup of materials are also **ferromagnetic**. The response of a ferromagnetic material to an external magnetic field is a very strong magnetization that can be orders of magnitude more than the external field itself. In ferromagnetic materials, not only do the magnetic moments of electrons not cancel out, but there are relatively large volumes of the material (called **domains**) where all the electrons have spins that are aligned with one another due to their mutual influence. This is responsible for a ferromagnetic material's strong response to an outside magnetic field.

When a ferromagnetic material comes under the influence of an external magnetic field, the domains already aligned in the direction of that field will start to grow, forcing the other domains to shrink to accommodate them. When this happens, the material is said to be magnetized, that is, it will

Magnetism vs. Gravity

We know that everything with mass exerts a gravitational tug on every other mass. A lamp across the room pulls at a cup of tea at your side, the Golden Gate Bridge pulls at Mount Everest, and the Sun pulls at the Earth as a whole. In all cases gravity pulls rather than pushes, and the gravitational force from most objects are swamped by the planets and Sun: the Earth's gravity pulls at both the teacup and lamp much more than they pull on each other, so they remain in place rather than flying across the room.

Magnetism, however, can either pull or push. As described elsewhere in the chapter, for most materials the attraction and repulsion cancels out at the level of atoms, leaving no net force. Some materials can temporarily have those forces out of local balance, and some can have them permanently out of local balance. These last materials are the objects we tend to call "magnets" in everyday speech.

temporarily act like a magnet itself. When the field is removed, the material will try to go back to its original state, but ferromagnetic materials have a tendency to stay magnetized to a certain extent—they often have some "memory" of past exposure to magnetic fields. Examples: nickel, cobalt, iron, and a selection of metals composed of rare earth elements. (There is another type of material known as *ferrimagnetic*, which can exhibit some properties of ferromagnetic materials but are distinguished from them by subtle differences in properties at the atomic level. These materials, like the mineral magnetite, were originally considered to be ferromagnetic.)

As noted, you can create a temporary magnet with some materials by placing them in an external magnetic field. For example, a bar magnet can be used to make something like a paperclip into a temporary magnet. If you use the magnet to pick up the paperclip, that paperclip will then have the electrons within it all lining up in response to the magnetic field of the magnet you are using. And now that first paperclip can pick up other paperclips. If you pull the magnet away from the paperclips, they will hang together for a short moment, then fall apart.

If iron placed in a magnetic field is only a temporary magnet, where do permanent magnets, like bar magnets and refrigerator magnets come from? Normal iron that you might find in nature will have small magnetic domains within it, but typically the domains do not all point in the same direction, and so over the whole of the piece of iron, the magnetic field essentially cancels out. The only naturally occurring material that is magnetic without an outside field is called **lodestone**, a certain variety of the iron-bearing mineral magnetite. This is the only natural magnet with the polarity and attracting ability that we would normally recognize as a true magnet. The commonly seen horseshoe-shaped magnet is a human invention. It is an iron alloy that is made to be a permanent magnet by subjecting it to a strong magnetic field during the forging process.

Table 8.1 Table of Materials That Can Be Magnetized, and Their Curie Temperatures

Material	Formula	Curie Temperature (K)
Cobalt	Co	1388
Iron	Fe	1043
Magnetite	Fe_3O_4	858
Nickel	Ni	627
Chromium oxide	CrO_2	386
Gadolinium	Gd	292
Europium oxide	EuO	69

Making a magnet first requires heating of the ferromagnetic material to the point that it loses any previous magnetic properties because of the thermal agitation. This temperature is different from one material to another, and is called the **Curie temperature** or **Curie point**. While the material, say iron, is above this temperature, a strong external magnetic field is engaged. As the iron cools, the thermal agitation slows, and the electrons all line up with the external magnetic field. Iron needs some additional influence, like work hardening through hammering or vibration, to "freeze" the domain/domains within the iron all in one direction. The final result is a permanent magnet that exhibits its own magnetic field with its own north and south magnetic poles. The only way to completely and utterly wipe out the magnetic field is to raise it once again above the Curie temperature.

Electromagnetism

Now that we have some idea where both electric and magnetic fields come from and what they can do, we can discuss how they are related. Every charged particle generates an electric field. If that electric field is moving or changing in some way (technically referred to as an oscillating or time-variant field), it will also produce a magnetic field. The same is true in reverse. An electron is a tiny magnet with a magnetic field. If that field oscillates, it will also produce an electric field. That means these fields really do not exist independently. They are bound together, creating and interacting with one another to form a united whole, a complete phenomenon known as **electromagnetism**.

Two simple experiments commonly undertaken can illustrate this concept. In the first experiment, a magnetic field can be generated from an electric field (called an **electromagnet**). A normal iron rod can be wrapped with a length of copper wire with the wire ends attached to the terminals of a battery. As the battery forces an electric current through the wire, the iron bar will become a temporary magnet which can be used like a common bar magnet to pick up paperclips or other similar objects.

In the second experiment, a moving magnetic field will generate an electric field. A length of copper wire can be coiled, and the ends of the wire attached to a light bulb. If a strong magnet is placed inside the coil and moved up and down, it will generate an electric field. This field will mobilize the electrons in the wire, which will generate electric current, and the light bulb will illuminate. When the magnet stops moving or the wire is disconnected, the bulb will go out.

A variation of the second experiment calls for leaving the magnet in place and moving the wire around it, instead. Imagine taking the length of copper and the light bulb, and moving them up and down while the magnet is stationary. It has exactly the same effect, creating an electric field that lights

Electricity, Magnetism, Light, and Space

It was realized by the 1830s that current flowing through a circuit acted like a magnet, attracting or repelling wires depending on the direction of current flow. The first dynamos were constructed soon after, leading to their use in power generation during the 1870s. The construction of these machines and the study of the physics that made them work is quantified in Maxwell's Equations, named for compiler James Clerk Maxwell, which are mathematical descriptions of how electrical and magnetic fields are generated and interact.

In the early 1860s, Maxwell calculated that the speed of electricity was equal to the speed of light and realized that the equations governing electricity and magnetism also described light. This revolutionized the thinking of physicists with respect to light. Further revolutions were to follow in the early twentieth century, however, as Albert Einstein recognized that Maxwell's Equations and the gravitational laws of Isaac Newton could not both be right. Einstein's intuition suggested that it was Newton's laws that needed updating, and the Theory of Relativity was the result.

the bulb. This is essentially the design of a **dynamo**, where current is generated in conducting material as it moves within a magnetic field.

PLANETS AS MAGNETS

Depending on the nature of the interior of a planet, it can act like a giant permanent magnet in space. A typical rocky planet can have an iron/nickel core, which is an excellent material to both create and respond to magnetic fields. This is a simplistic view, however, since a planet is far too big, and the processes are far too complicated for the core of any world to be a perfect, solid permanent magnet. Actually, something quite different is going on.

Origins of Planetary Magnetic Fields

The origin of the Earth's powerful and durable magnetic field remained a mystery for a long time. Part of the mystery is that the interior of the Earth is hotter than the Curie temperature for iron, so no permanent magnet should be sustainable. Another issue is that magnetic fields within a conducting body undergo what is called **ohmic decay**, and for the Earth this would happen quite quickly. So another phenomenon must be in effect, maintaining the magnetic field.

The complexities surrounding the generation of Earth's magnetic field are still not completely understood, but it is now considered likely that the field is the result of a dynamo acting inside the Earth. This dynamo works because the Earth has a liquid outer core of conducting material to **convect**, and because the Earth is rotating rapidly. Earth's rotation allows the convecting material and subsequent fields to distort and align approximately north to south under the influence of the **Coriolis force**. The electric fields serve to produce and then reinforce magnetic fields and allow for yet stronger electric fields. The end result for the Earth is a strong magnetic field with a north and south magnetic **dipole**.

Not all planets have the same form of magnetic field. To have a field like the Earth's, a planet must have some liquid conducting material in its interior, that liquid has to be convecting, and the planet needs to be rotating quickly enough to distort the convecting areas. Mercury has a global magnetic field similar to the Earth's, but it is only about one percent as strong. Still, such a global field implies that Mercury also has a liquid, convecting region of its core. A planetary body that does not have a conducting, convecting region inside can still have some kind of magnetic field. Mars does not have a global field, but instead there are regions of the crust that are magnetized, allowing patchy areas on the planet's surface to create local magnetic fields.

Magnetic fields are not static and unchanging. A planet can even lose its magnetic field if it cools down enough so that the liquid regions within it solidify, or if its rotation is dramatically slowed. There is evidence that Mars had a more substantial magnetic field in its past, but it must have been billions of years ago.

Directional Information

On the Earth, our magnetic field has been put to practical use, the foremost being that it allows for the existence of the compass. As with any conductor in a magnetic field, the pointing arrow of a compass is redirected by the Earth's strong field. Compasses are designed so that the arrow points north, more or less. As you may know, compasses do not literally point to the north pole as defined by the Earth's axis of rotation. Instead every compass needle points to *magnetic north* by definition. This can differ from the actual north magnetic pole, which is the place at which the magnetic field lines of the Earth emerge in the northern hemisphere perpendicular to the planet's surface. This difference is due to the fact that a compass needle responds to the local magnetic field, which varies over the planet. And in turn, the north magnetic pole (and south magnetic pole) are not tied to the surface, but can wander. The dynamic, liquid core of the planet has motion independent of the material above it, and so does the resultant magnetic field. Right now these two "north poles" differ by about 10 degrees.

A planet without a global magnetic field may still have localized magnetization of rocks that could influence a compass needle. In that case, the needle might point in any direction, depending on the location of the rocks in question. Not very useful for navigating the world on a global scale, but possibly good as a means of finding your way over short distances near those deposits. More likely, instrumentation would be designed to find

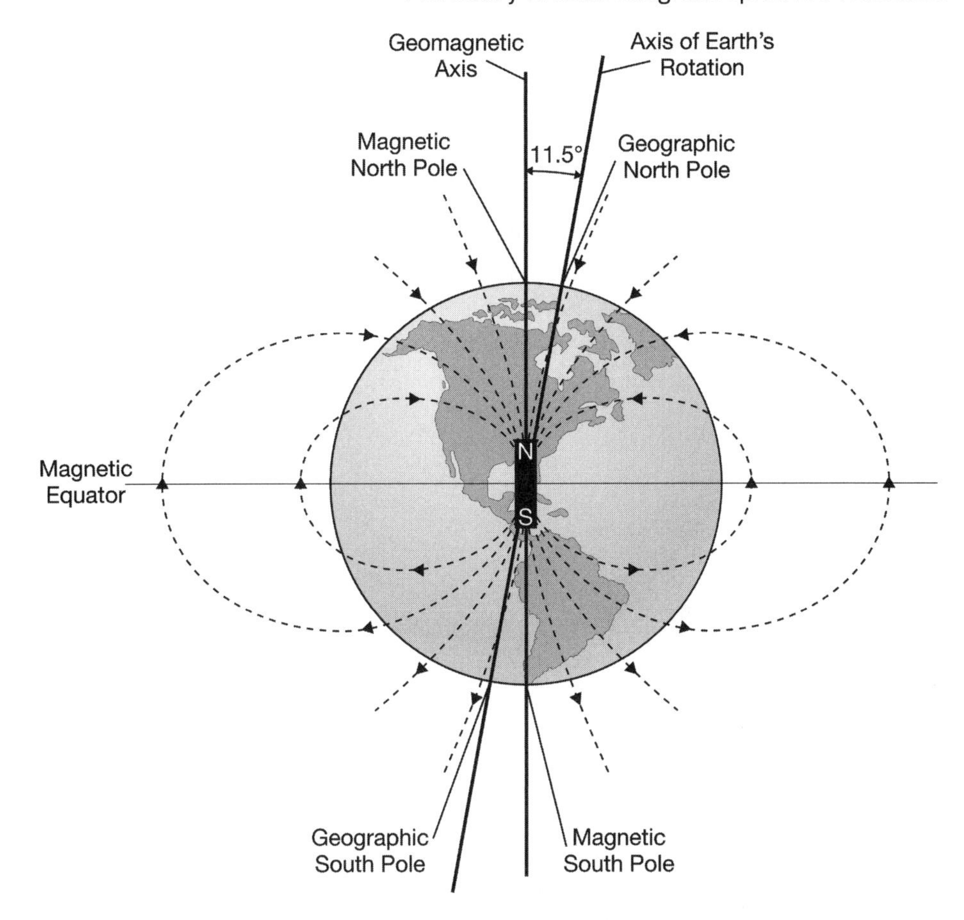

Figure 8.1 The magnetic field of the Earth is generated in its core. The axis of the Earth's magnetic field is not aligned with the rotation axis, which is why compasses do not point exactly north. The relative positions of the magnetic north pole and north pole can change with time, and from time to time the polarity of the magnetic fields switch, with the north and south poles swapping positions.

those deposits, and navigation would be conducted more as it now is on Earth, by satellites and GPS (Global Positioning Systems).

Magnetic Influence on Rocks

As we already discussed, external magnetic fields can create a magnetic response in ferromagnetic materials that will remain to some extent after the field has been removed. If the material was above its Curie temperature at some point, and cooled in the presence of this external field, it may retain an excellent memory of that field.

This is a powerful phenomenon in planetary science. A planet can imprint its magnetic signature on its own rocks. Mars, as noted, does not have a global magnetic field. But if the dynamo theory for Earth is correct, it implies that a very young Mars would have had a similar, but less powerful dynamo. Mars, being much smaller than the Earth, has apparently cooled to the point that the dynamo no longer exists. Even so, rocks on the surface of Mars retain some memory (remnant magnetism or **remanence**) of the past field. The patchy magnetized regions of its planetary crust are those areas with the proper minerals to have recorded and remembered the old Martian magnetic field for as much as four billion years. Not only does this mean scientists have a method for studying magnetic fields that existed billions of years in the past, it helps them isolate particular minerals within the crust of Mars right now.

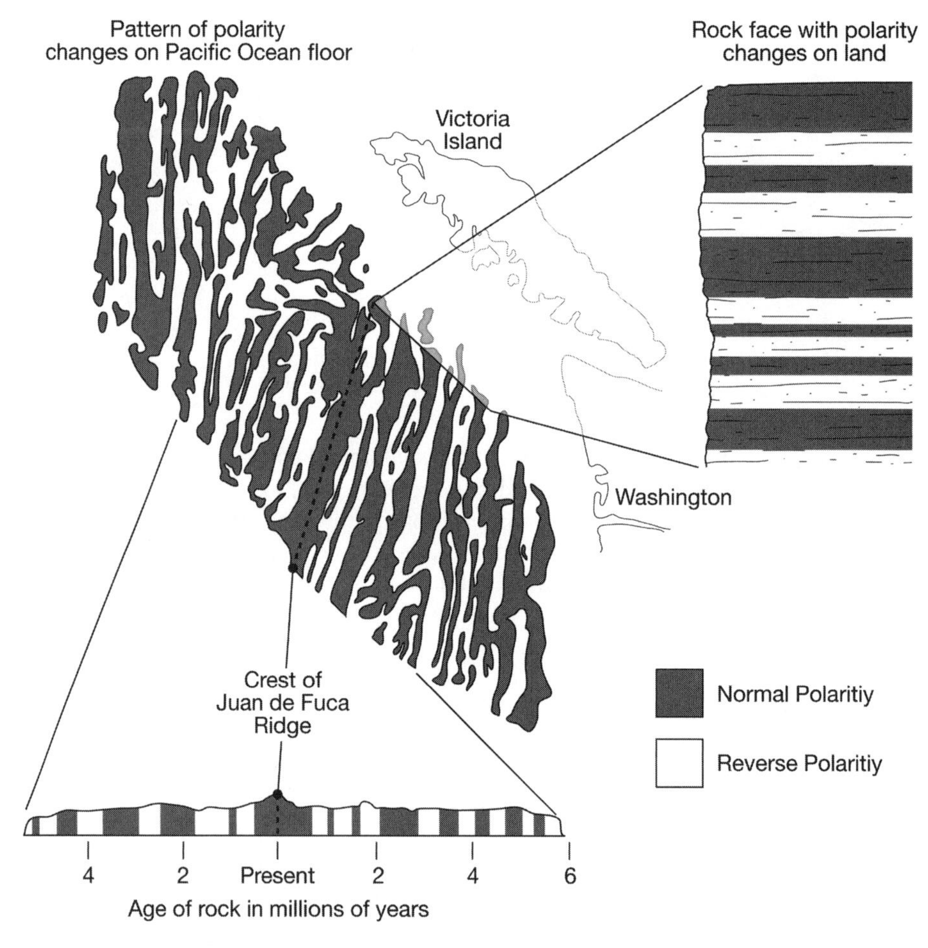

Figure 8.2 As rocks cool, some of their minerals can retain traces of any magnetic fields present as they pass through the Curie temperature. Rocks on the ocean floor form at ridges, with older rocks being found further from the ridge. The polarity seen in the magnetic fields recorded in the ocean floor rocks can be used to determine an age. The diagram above, from the USGS, shows how geologists interpret the magnetic stripes on these rocks and connect them to rocks on the land and to their ages.

Some types of meteorites are nearly as old as the solar system itself. Models of solar system and planet formation originally assumed that the building blocks of planets were cold lumps of rock. But scientists have found magnetic signatures in ancient meteorites suggesting that the parent bodies or asteroids that they came from, billions of years ago, may have not only been melted, but melted enough to create their own dynamos! If this theory is correct, it has far-reaching implications for all theories of how the inner planets, including Earth, initially formed.

One of the most compelling proofs for the theory of plate tectonics was found on Earth's ocean floor. The theory required that new crust was being created as old crust was subducted and destroyed; an idea that many scientists initially found preposterous. But the sites where new crust was being extruded were located at the bottom of the ocean floor, in mid-ocean ridges. As molten crustal rocks were forced upward, they cooled and retained memory of the Earth's magnetic field. Specifically, they remembered the direction of the field. It had already been shown that occasionally the Earth's magnetic field would flip, and what was once magnetic north would become magnetic south (and vice versa). Looking at the magnetic signature of the ocean floor revealed magnetic bands parallel to the ridge on each side that were mirror images of one another. This was strong evidence in favor of new crust coming up from the ridge and being forced outward to either side, spreading the sea floor. And it allowed for an estimation of the timing of the formation of the new crust, as well. Plate tectonics and sea floor spreading are, of course, processes that continue today.

A PLANETARY MAGNETOSPHERE

Just as with defining the atmosphere for a planet, defining **magnetosphere** can be a challenge. It is sometimes used synonymously with "planetary magnetic field" or "region under the influence of a dipole," but this is not quite accurate. The magnetosphere of a planet results from the interactions between its magnetic field and the solar wind, including the potential involvement of a planetary atmosphere as well as the larger **Interplanetary Magnetic Field** (IMF) that is always in play throughout the solar system.

There is a region around the Earth, a sphere of influence (not actually shaped like a sphere), where the Earth's magnetic field has a dramatic effect on the movement of charged particles. Charged particles are impinging on this field at all times, most of them a part of the solar wind streaming from the Sun. Where the solar wind and the magnetic field meet forms the boundary of the planetary magnetosphere. So a magnetosphere is that area around a planet or other body where phenomena, like particle interactions, are dominated by the body's magnetic field. The shape isn't an actual sphere because the solar wind pushes on the magnetosphere and forces it into a teardrop shape, with its tail "downwind" of the solar wind. And as one would expect from our previous discussion, the sort of particles that can be diverted by a planetary magnetic field are ones that have charge. Something like an electron, but also atoms, if the atoms have been **ionized**. An ion is an atom that is missing one or more electrons. Such ions will no longer be in charge balance, and so will also fall under the influence of magnetic fields.

Figure 8.3 is a diagram of the planetary magnetosphere we know the most about, the Earth's. For the Earth, the magnetosphere extends about 10 earth radii (roughly 60–65,000 km) outward toward the Sun, and 200 or more Earth radii (roughly 1.2–1.3 million kilometers) down the **magnetotail**. Ahead of the edge of the magnetosphere is the **bow shock** region, where streaming particles moving very fast are suddenly decelerated. The bow shock is named for the wave seen in front of the bow of a fast-moving ship. Most of the solar wind is pushed sideways, away from the

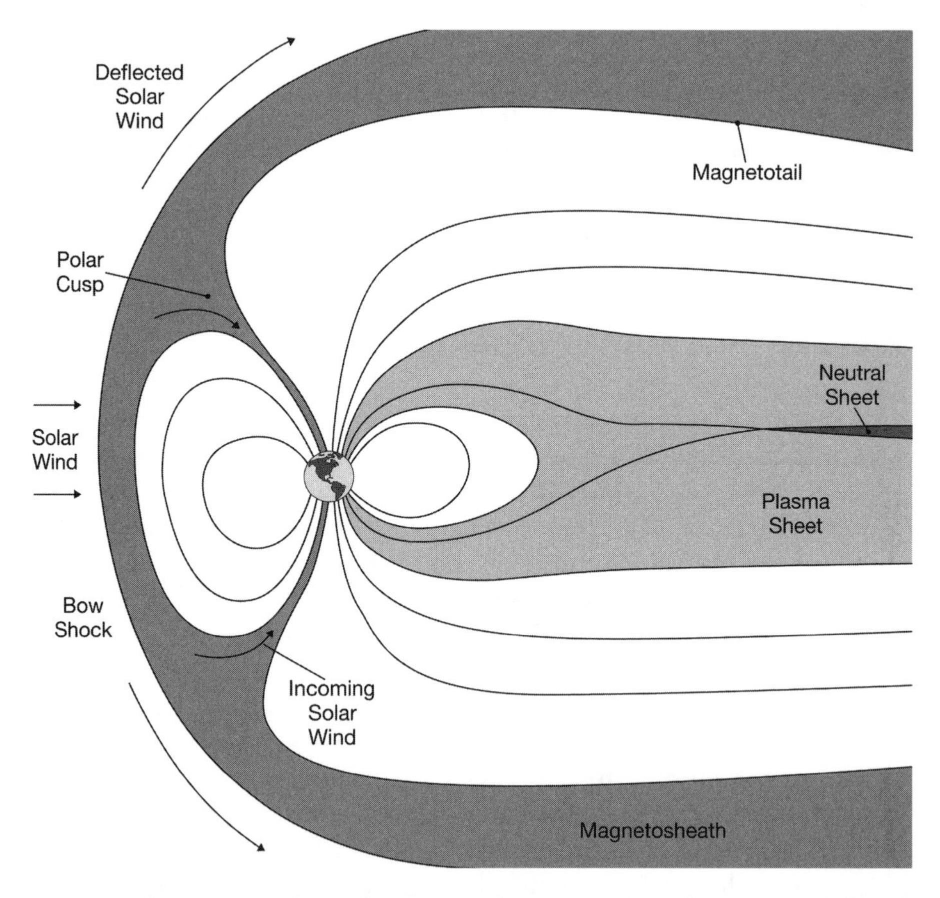

Figure 8.3 The magnetosphere of a planet is the region where the magnetic fields of the Sun and the planet interact. Despite its name it is not spherical in shape, instead with a shape that is changeable dependent upon the output of the Sun.

Earth, flowing around and down the magnetotail. The "surface" of the contact of particles down the magnetotail is called the **magnetosheath**. The magnetosphere is dynamic and constantly changing in response to solar events. The solar wind is not a perfectly even, calm flow. It varies in strength and intensity, and the magnetosphere of the Earth constantly changes with it.

LIFE IN THE STELLAR NEIGHBORHOOD

In many ways, the Earth and the other plants can be said to exist within the atmosphere of the Sun. All these bodies are constantly bathed by particles and energy and subject to the Sun's stormy moods. Separate from the scientific curiosity about the nature and type of the interactions, we need to understand how planets interact with their central star because it has direct ramifications for the development and persistence of life. Particles and energy from the solar wind can alter atmospheres, change surface rocks, as well as bombard and damage living creatures, and in our case, their sensitive electronic devices.

General Solar Activity

The Sun is not a quiet place. It is a massive ball of hot **plasma** with a fusion reactor at its core. It generates a powerful magnetic field that entrains highly energetic particles as they stream away from the solar surface, and that surface is a churning mass of plasma, constantly moving and convecting. The Sun emits electromagnetic radiation in all wavelengths, including visible, ultraviolet, and gamma rays. The magnetic field surrounding the Sun can suffer a sudden change that creates a solar storm. Such storms are often accompanied by a **solar flare** or a **coronal mass ejection (CME)** sending a blast of charged particles and high-energy radiation into the solar system. If we are in the path of one of these events it could create a **geomagnetic storm**; distorting our magnetosphere with a host of effects.

Effect on Atmospheres

A planet without a significant magnetic field has little protection from the solar wind. If such a planet has an atmosphere, the solar wind will strip some of the atmosphere away over time. How dramatic this effect is varies from one planet to another. But both Mars and Venus have lost (and are still losing) atmosphere to the solar wind, and probably lost some, if not most of their water that way, too.

The Earth's magnetic field helps protect us from losing a great deal of our atmosphere to the solar wind. But the solar wind still induces changes in our atmosphere with important consequences. There are changes to the temperature of the upper atmosphere and its electrical and chemical properties, as well as alteration of the ozone layer and some minor atmospheric stripping. Naturally, any loss of atmosphere is a source of concern. Scientists need to continue to study the mechanisms for all these changes so we have a solid understanding of the stability of our atmosphere and environment.

Our magnetic field does not block each and every particle from the Sun, some still leak through, and along with X-rays cause the Earth's upper atmosphere to heat. The resulting expansion of the atmosphere due to this heating puts substantial extra drag on satellites in low Earth orbit. Satellites in such orbits must be equipped with the means to give themselves a boost every now and then, or they will eventually see their orbits degrade to the point that they re-enter the atmosphere and fall back to the ground. Attempting to place a satellite into orbit during a solar storm with its associated extra drag could result in the orbit ending up too low for the satellite to be useful.

The Sun's ultraviolet radiation also causes some ionization in Earth's upper atmosphere, which is partially responsible for the existence of the layer in our atmosphere called the **ionosphere**. Particles coming from the solar wind, as well as some from the ionosphere provide a source of plasma that becomes trapped within our magnetosphere.

Effects in Space and on the Ground

Of course, having a satellite in a lower orbit than expected is typically better than having it fail completely or be destroyed. But failure and destruction can also be the consequences of a geomagnetic storm. Satellites have electronic equipment that is highly sensitive to radiation, just as on a smaller scale we are advised to keep our credit cards and computer hard drives away from strong magnets. A sudden blast of charged particles, along with interactions between the solar wind and the magnetosphere's trapped plasma can dramatically change the radiation environment near Earth. This is dangerous for satellites, as well as any astronauts that might be in space at the time. If solar storms can be predicted, satellites can be instructed to enter protective "safe modes" and astronauts can be certain to take steps to shield themselves until the storm passes.

A geomagnetic storm can even create havoc with power systems all the way down on the surface of the planet. Such a storm can generate currents in power lines that can cause power plants to fail, damage to computers and other electronic equipment, and disruption of communications in telephones, televisions, and radios with bursts of static.

The Solar Storm of 1859

The most powerful geomagnetic storm ever recorded occurred at the dawn of the electrical age. For several days at the end of August and start of September 1859, the northern lights were seen much further south than usual, as far south as Hawaii. Townsfolk throughout North America unfamiliar with auroras sent fire brigades rushing through the countryside to find the fires that they thought were responsible for the reddish glows they saw. Telegraph stations, the modern technological wonders of the era, were destroyed as the changing magnetic fields caused by the storm-induced currents in the wires, overloaded them and caused real fires.

The storm was an important event in our understanding of the Sun's interactions with the Earth. The first instruments capable of measuring the strength of magnetic fields were only recently installed at various sites and showed that the storms had a magnetic component. A British astronomer observed large solar flares in the days preceding the most intense part of the storm, providing evidence that the Sun was responsible. A network of aurora observations by oceangoing ships provided a wealth of data. The information obtained during the 1859 storm, even from the primitive instruments available, continues to be analyzed to help us understand what might happen the next time such a large event occurs, and to plan for mitigating its effects.

Aurorae

Earth-Sun interactions are certainly not all bad. Despite the dangerous situations that can arise on or near Earth, these interactions are also responsible for some rather beautiful events. A magnetosphere cannot stop every charged particle that impinges on it. And as some of these particles collide with atmospheric molecules, they can produce amazing and spectacular phenomena called *aurorae*, otherwise known as the northern (aurora borealis) and southern lights (aurora australis).

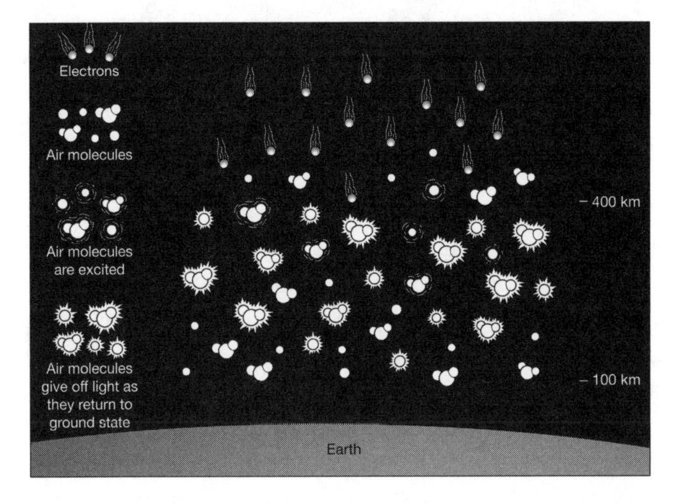

Figure 8.4 The aurora, or Northern (and Southern) Lights, occur high in the atmosphere, as charged particles from the Sun interact with atmospheric molecules.

Figures 8.5 From the ground (top panel) and from space (bottom panel), the aurora are an impressive spectacle, obvious evidence of the interaction between the Sun and the Earth.

It is known that aurorae can occur in those regions where a planet's magnetic field funnels charged particles down to interact with the atmosphere. Therefore an aurora requires both an atmosphere and some kind of magnetic field. On Earth and the Giant Planets, aurorae occur around the polar regions because that is where the particles are directed by these planets' strong magnetic fields. However, a planet can still have aurorae without such a strong, planet-wide field. Rocky planets like Mars can have an aurora form over any area of the planet where the underlying rocks of the crust are themselves magnetic. These are far less powerful and dynamic, but they have been observed nonetheless.

A Word About Life

It may be that a global magnetic field is required for life to arise and persist on any world. The solar wind, and especially energetic solar storms, deplete planets of their atmospheres and water. They bathe the surface of planets in charged particles that can cause damage to life forms. Once an atmosphere is sufficiently depleted, there is no protection from any kind of radiation. As the search for life on worlds in other star systems continues, the existence of a magnetic field might point the way to the most likely places to look. Certainly our own survival on Earth has been dramatically impacted by the existence of our magnetic field. And without it, our ability to create and operate sensitive electronic equipment would be seriously compromised.

CONCLUSION

Although invisible to the eye, magnetic fields are pervasive in our environment. The Earth is thought to generate its planet-wide field deep in a liquid metallic core, which effectively acts as a dynamo. Mercury also has a planet-wide magnetic field, while the Moon, Mars, and Venus do not appear to have such fields. The presence (or absence) of a magnetic field generated by a planet is seen as evidence for (or against) a liquid core. A planet's magnetic field interacts with the magnetic field generated by the Sun, with results seen from the ground as aurorae. The changing solar field can push on the planetary field, with the fields changing shape in response. Rocks that cool in the presence of a magnetic field can retain signatures of those fields, depending on the minerals they contain. This has allowed us to study ancient magnetic fields here on the Earth, as well as study evidence for ancient fields on the Moon and Mars.

FOR MORE INFORMATION

A discussion of the difference between the Earth's rotation and magnetic poles and the way the Earth's magnetic field has changed and continues to change with time is found at http://science.nasa.gov/headlines/y2003/29dec_magneticfield.htm.

An in-depth report on the giant solar magnetic storm of 1859 is included at http://www.space.com/scienceastronomy/mystery_monday_031027.html.

A general overview of space physics, including magnetism is presented at http://www.phy6.org/readfirst.htm. Topics closely tied to the material in this chapter can be found at the pages http://istp.gsfc.nasa.gov/earthmag/inducemg.htm and http://istp.gsfc.nasa.gov/earthmag/dmglist.htm.

The National High Magnetic Field Laboratory hosts a wide array of information about magnetism at http://www.magnet.fsu.edu/education/, with a set of pages concerning electromagnetism specifically at http://www.magnet.fsu.edu/education/tutorials/electricitymagnetism.html.

9

The Moon: Planetary Rosetta Stone

INTRODUCTION

The Rosetta Stone, discovered in 1799, was the key to finally unlocking the code to the hieroglyphic writing of ancient Egypt. The stone has the exact same passage of text presented in classical Greek, Demotic script, and hieroglyphic writing. The term "Rosetta Stone" has idiomatically come to mean the critical resource needed to solve a difficult puzzle.

The Moon is the "Rosetta Stone" for understanding the basic chronology of the formation and evolution of the terrestrial planets. The surface of the Moon is ancient, and unlike the Earth, it retains a record of events from the early days of the solar system. Our ability to use the Moon as our key comes from that fact that we have been there in person and brought back samples of rocks and lunar soil. Dating those samples with radiometric techniques means we can calibrate the lunar cratering record, and then apply that knowledge to the cratering records on the other inner planets.

HISTORICAL BACKGROUND AND EXPLORATION

The Moon has long been important to humans. It was worshipped by ancient civilizations and figures deeply in folklore across the world. It was thought to have particular power over people, a concept still present in the English words "lunacy" and "moonstruck." It was also used pragmatically by ancient peoples as a means of timekeeping. The word "month" is derived from the Moon, and the Jewish and Muslim calendars are tied to the lunar phases.

The Moon keeps one side to the Earth, called the near side, as discussed below. Telescopic observations of the near side led astronomers of the 1600s to divide the Moon into "land" (terra) and "sea" (mare) and name the major features; names which are still used today. By the 1800s, fairly detailed maps of the Moon were available, though many important science questions remained unanswered at the dawn of the Space Age.

The first spacecraft to reach the Moon was the Soviet *Luna 2* mission, which crashed into the Moon in September 1959, less than two years after *Sputnik* became the first satellite to orbit the Earth. A series of spacecraft visits followed in rapid succession as part of the "Space Race," as the United States and Soviet Union vied for the first piloted landing on and return from the Moon. Some notable missions in the Space Race included *Luna 3*, first spacecraft to return pictures of the Moon's far side, the *Ranger* series, which were American probes designed to crash into the Moon, the *Surveyor* series, which were soft landing robots, and the *Lunar Orbiter* series, which returned detailed pictures of the lunar surface in order to pick landing sites. The Soviet space program long held the lead over the United States, with *Lunas 9* and *10* serving as the first soft lander and the first lunar satellite, both in 1966. However, the *Apollo* program first put humans on the Moon in 1969, and continued through 1972, ending the Space Race and returning over 380 kg of priceless lunar samples. The Soviet program continued for several more years and included two rovers and sample returns of more than 300 g of material from robotic missions.

After the final *Luna* sample return in 1976, a long hiatus in lunar exploration followed. The next successful mission to the Moon was *Clementine* in 1994, a NASA orbiter which mapped the Moon at higher resolution than was previously available. *Lunar Prospector* in 1998 was the next NASA mission to the Moon. The early 2000s saw another space race of sorts, as the European Space Agency, Japan, India, and China all sent orbiters to the Moon with imagers, spectrometers, and radar. The United States plans further lunar missions as well. While the last astronaut left the Moon more than 30 years ago, plans for a permanent lunar base are being considered by the United States and China, among other nations.

GEOLOGIC HISTORY

Along with the rest of the planets in the inner solar system, the Moon formed about 4.5 billion years ago. The body was being constantly bombarded by the material remaining from the formation of the inner solar system as a whole, which served as the very last stage of planetary accretion. The very early Moon was probably covered by an extensive **magma ocean** whose suspected depth varies between models. Although it must have been at least 100 km deep, it was probably more like 500 km, and some models find the entire Moon was melted. Ancient lunar zircons have been dated to approximately 4.4 billion years ago, which suggests that the bulk of the magma ocean had **crystallized** by that time, although younger zircons indicate that some crystallization would have taken place for another several hundred million years. During this solidification and crystallization period, the denser minerals such as **olivine** and **pyroxene** sank to the bottom of the magma ocean, while the lightest minerals such as **plagioclase** "floated" to the top. This created a crust of **anorthositic** plagioclase **feldspar**, represented today on the Moon by the anorthositic highlands. The last portion of the early magma ocean to crystallize was concentrated in potassium (K), rare earth elements (REE), and phosphorus (P) and so has been called KREEP. These magmas found their way to the near side where they formed the KREEP-rich suite of rocks near Imbrium and part of Procellarum basins.

There is a distinct difference in the europium (Eu) abundance between the plagioclase rocks of the **lunar highlands** and the mare basalts. The amount of Eu is anomalously high in the highlands rocks (a positive **europium anomaly**) and anomalously low (a negative europium anomaly) in the mare basalts. Europium has a tendency to prefer going into plagioclase during crystallization rather than into other minerals. This Eu anomaly is additional evidence that the highlands and the mare were originally part of the same general source, with the plagioclase crystallizing out first and "taking" most of the europium with it.

The surface had solidified by 4.1 to 4.2 billion years ago, and at least one of the major basins on the lunar surface, Nectaris, may date to that time interval (or it may be younger). There is some speculation that there was a sudden increase in basin-forming impacts 3.9–3.8 billion years ago (see the section on the lunar cratering record, below). The period of more intense bombardment of impactors was followed by widespread volcanic activity from 3.8 to 3 billion years ago. This included the extrusion of massive flood basalts of high-iron abundances that often pooled in the topographic lows that were created by the giant impact basins such as Imbrium and Serenitatis. While most activity seems to have happened in this 3.8–3 period, some basalts have been dated as early as 4.2 billion years ago.

After 3 billion years ago, the Moon became largely quiescent, with some minor volcanic activity, including some mare basalt extrusion possibly dating to less than 2 billion years ago (or even one billion years ago depending on the calibration of crater count data). The intensity of impactor bombardment continued to fall off. Today the geologic activity of the Moon is almost entirely limited to the occasional creation of impact craters (the most recent large crater was Tycho, formed 100 million years ago), a constant stream of tiny micrometeorites impinging the surface, small moonquakes, and the results of surface exposure to the solar wind.

PLANETARY INTERIOR

The Moon has a crust, mantle and relatively small core that is less than 500 km in radius. Certainly the Moon's mass and low bulk density (3.35 cc) would suggest a small core. It is possible that the outer core is molten, but insufficient seismic data exist to be certain. However, small perturbations in the Moon's orbit would support the idea of a molten outer core. The lunar crust is about 50 km thick, but could range from 30 to 70 km depending on the location and error in measurement. The far side of the Moon has a thicker crust than the near side by as much as 15 km. The interior of the Moon appears to have "cooled" off completely, such that the world is no longer geologically active.

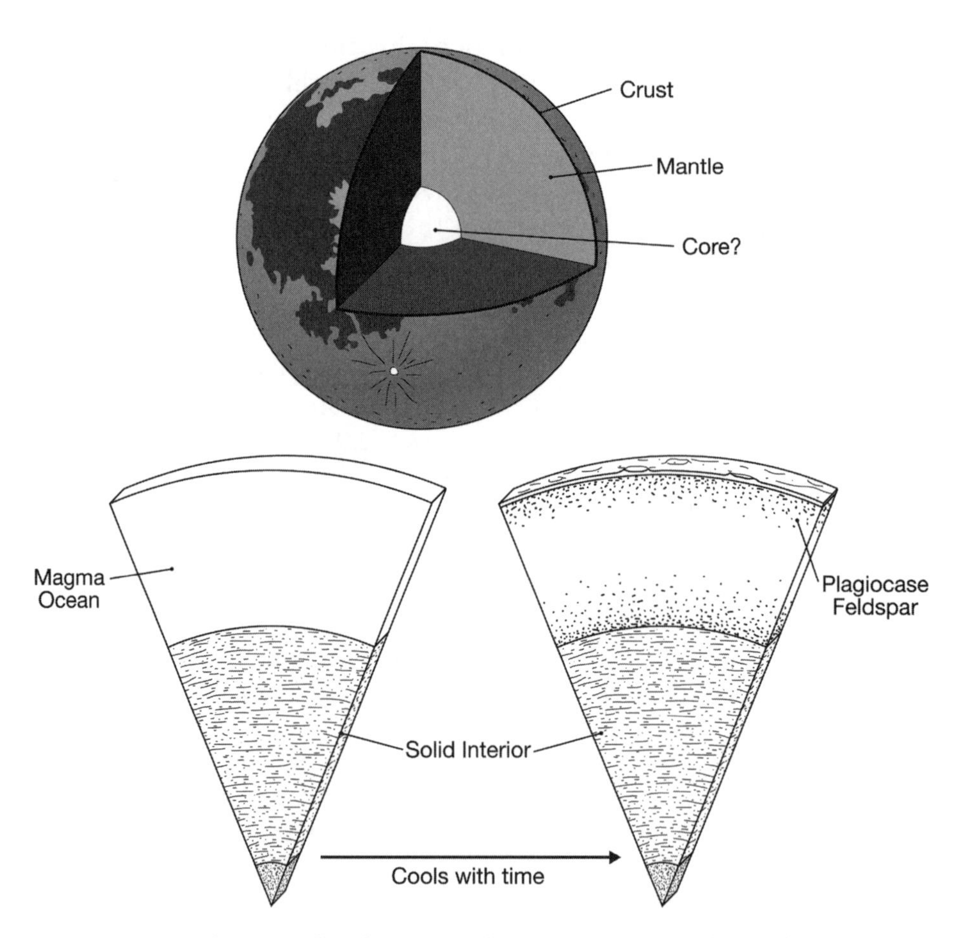

Figure 9.1 Near the time of its formation, the Moon was covered by a deep ocean of molten rock, or magma. As the magma cooled, minerals began to crystallize, with some sinking and some (like plagiocase feldspar) rising. Eventually, the Moon cooled completely, resulting in the structure we find today.

ATMOSPHERE

The Moon is often referred to as an "airless" body, and some would describe it has having no atmosphere at all. However there are particles and ions generated from various sources that exist above the surface level, and these do constitute a very tenuous form of atmosphere. These particles must be continuously replenished, as they are constantly being lost to space or reacting with the lunar surface.

Gasses are still being released by the interior of the body. Some, like argon and radon, are produced by natural radioactive decay. The surface of the Moon is not protected by a global magnetic field, and is therefore open to bombardment from charged particles from the Sun. The particles impinge on the surface, and interact with rocks and soil, sometimes combining and making new molecules or **sputtering** particles outward. The surface is also constantly being bombarded by small bits of dust from space (micrometeorites). The result of all these processes can be additional gasses added to the atmosphere, such as hydrogen and helium, as well as small amounts of other elements like sodium and potassium.

SURFACE FEATURES

Looking up at the Moon from Earth, one can easily identify the two major geologic provinces on the lunar surface. These are the bright, rough **highlands** areas, and the dark, smooth volcanic plains called **maria** (or in the singular, **mare**). It is also easy to identify the next major feature of the lunar surface, the fact that it is covered with impact craters of all sizes. Other major features can be more difficult to identify from Earth, since we only ever get to see one side of the Moon. Since the Moon is locked in **synchronous** rotation with the Earth, only the **near side** of the Moon faces us. The **far side** of the Moon always points away from Earth. The far side is also incorrectly referred to as the "dark side" of the Moon. But the far side is not dark, and is in sunlight as often as the near side. When the Moon is in the **new moon** phase, and the near side is dark, it is the far side that is in the full light of day from the sun.

Using spacecraft data, however, we can compare both the near and far sides, and then another major geologic feature of the Moon becomes apparent. Almost all the maria are on the near side. The reason for this is not entirely clear, but there are other key differences between the near and far sides that may be related. The lunar crust is thicker on the far side than the near side. It is also speculated that the near side has a greater concentration of radioactive elements. All of the KREEP basalts are on the near side. In addition, the major-basin forming impacts may have allowed for mantle uplift underneath them. Evidence for this is found in the gravity data for the Moon, which shows large gravity anomalies or

The Dry Mare

One of the big surprises from the *Apollo* program and the samples that were returned by it was the lack of water or any evidence there had ever been water on or in the Moon. This has been an important constraint on theories of lunar formation and has also been seen as a serious issue with respect to establishing a base on the Moon.

Things began to look somewhat better for being wetter in the 1990s as the *Clementine* spacecraft found some evidence that ice could be lurking inside craters near the lunar poles, thought to have been brought there by cometary impacts. Due to their location and depth, some craters contain spots that never see the Sun. These spots can get extremely cold; cold enough to keep water ice stable. The last several decades saw additional work with radar and mass spectrometers attempt to confirm and pinpoint regions that could have ice either buried below a small amount of regolith or at the surface. Plans for lunar bases tended to favor areas near the poles where this ice could potentially be turned into usable water.

Finally, a reanalysis of *Apollo* samples suggests that the Moon *could* have formed with water after all. These very precise new measurements are still being interpreted, but open the door to water being present in the lunar interior today, and suggesting that additional changes may be necessary to our current theories of lunar formation.

mascons in the basin regions that are larger than can be accounted for by infilling of maria.

Tectonic Activity

Although far more tectonically stable than the Earth, the Moon still undergoes minor tectonic modification, and shows evidence for past tectonic processes. Evidence for past tectonics includes graben formed along the

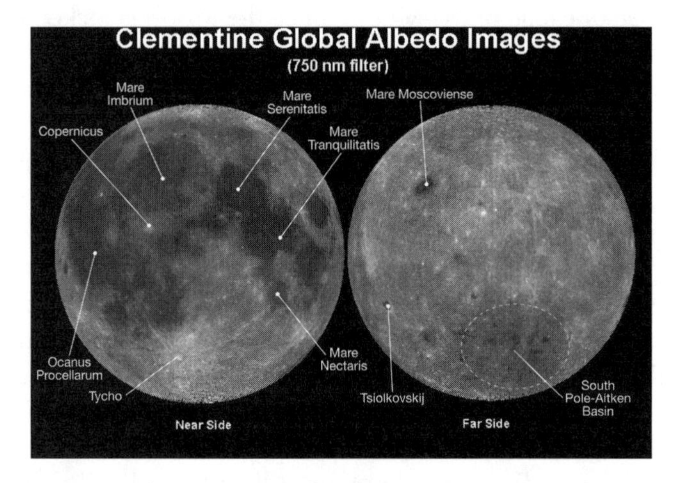

Figure 9.2 The near (left) and (far) sides of the Moon, with major features labeled. Places on the near side received their names starting in the seventeenth century, while the far side remained unseen until 1959. Because Soviet spacecraft were the first to image the far side, many of the far side names are Russian or Russian-related.

margins of certain mare inside of major impact basins. **Wrinkle ridges** also found within the mare are formed by compressional stresses.

The Moon does undergo "moonquakes" but these are most often less than a magnitude of two on the Richter scale. These are likely to be a result of gravitational tidal interactions between the Earth and Moon, although others are generated by impact events.

The constant bombardment of the surface by impactors big and small means that the lunar surface is always in the process of being reworked. One result is continuous downslope movement or mass wasting from mountains and crater rims. This is a source of slow but ongoing modification of craters, causing softening of the crater rims and slumping inside crater interiors.

Volcanic Structures

The Moon possesses vast plains of flood basalts (the maria), small volcanic domes, shield volcanoes, and volcanic rilles. Some of the components of the lunar soil are **pyroclastic** in origin, such as small glassy spherules. These may have been formed as molten rock fountained out of volcanic vents and solidified quickly into droplets before falling back to the lunar surface. Volcanism appears to have been the most influential process shaping the

Figure 9.3 This wrinkle ridge, photographed by the crew of *Apollo 15*, is interpreted as evidence of compressional stress within mare basalts, which shrunk as they cooled.

lunar surface other than impact cratering. The domes and shield volcanoes appear to be source regions for lava or pyroclastic eruptions. The sinuous and arcuate **rilles** are ancient lava channels that once carried lava away from eruption vents.

Cratering Record

Impact cratering has been the dominant process shaping the surface of the Moon. And while it was far more dramatic in the past, it still continues today, unlike many other processes that ended when the planet became geologically quiescent. In general terms, the early Moon was impacted by a large number of massive objects that were probably left over from the formation of the Moon and planets in the inner solar system. This would have included some comets, which may have been responsible for trace amounts of water on the Moon, and brought some water to the Earth, as well. Since the accretion of the planets, the number of impactors available to strike planetary objects has been going down, as they have mostly been swept up by the planets in their paths, been thrown into the Sun, or sent right out of the solar system. There are still rocky objects and comets that impact the Moon and planets today, but this happens much less frequently than it did. And this is very good news for life on Earth.

The largest and earliest impacts into the lunar surface formed the massive impact basins including Imbrium, Crisium, Serenitatis and Nectaris. Scientists have attempted to date each of these impacts radiometrically, although we do not have rock samples that can be unambiguously tied to all basins. However, scientists have generated ages for most of the basins either radiometrically or by counting the craters superposed on mare infill. Earliest lunar history is divided into the Pre-Nectarian and Nectarian time frames, with the age of the Nectaris basin defining the division. This is a bit ambiguous, however. While the first dates for Nectaris placed its age nearer to 3.9 billion years ago, it has also been dated closer to 4.1 billion years ago, or more. The division between the age of the Nectaris basin and Imbrium basins (one of the youngest basins) defines the extent of the Nectarian period. So this period is either quite short, from 3.9 billion years, to the age of Imbrium at around 3.85 billion years. Or it is rather longer, perhaps as long as 4.2 to 3.85 billion years.

The age of all the basins, but perhaps more specifically Nectaris, is of critical importance in understanding the early bombardment history of the Moon. If Nectaris is young, then almost all of the major basins formed in a very short period of time, relatively speaking, between 3.9 and 3.85 billion years ago. Such a massive influx of impactors would either have been a tail off of the late heavy bombardment of the body, or may have represented a discrete and sudden increase in the impact cratering flux called the

terminal lunar cataclysm. Planetary scientists continue to try to refine ages for the basins in order to see if the cataclysm hypothesis is borne out by the evidence. At this time, even if Nectaris is older, there still seem to be a very large number of basins formed in a short interval. If a cataclysm did occur on the Moon, it has wide-reaching implications for the formation of the Moon and the solar system as a whole. Scientists need to define where such a body of impactors might have been sequestered for 500 million years to be available for such a sudden influx into the Moon, and also what may have triggered their sudden move into Moon-crossing orbits. Newer models for solar system formation (i.e., the Nice model) might be able to explain this cataclysm if gravitational interactions between Jupiter and Saturn drove those planets to a **resonance** that flung outer solar system icy bodies into the inner solar system.

The Imbrian period ended about 3.2 billion years ago, and the Eratosthenian period began. During this period, which lasted until about 1.1 billion years ago, the cratering flux continued to fall off. The final era of lunar cratering history, the one we are in now, is the Copernican, 1.1 billion years ago to the present. This time period is named for the crater Copernicus, which is about 100 km across and 810 million years old. A number of the craters formed during this era are young enough to still be surrounded by bright rays of ejecta. Although, the larger craters preferentially retain their rays longer than the smaller ones. Also within this cratering era the crater Tycho was formed; similar in size to Copernicus, but only

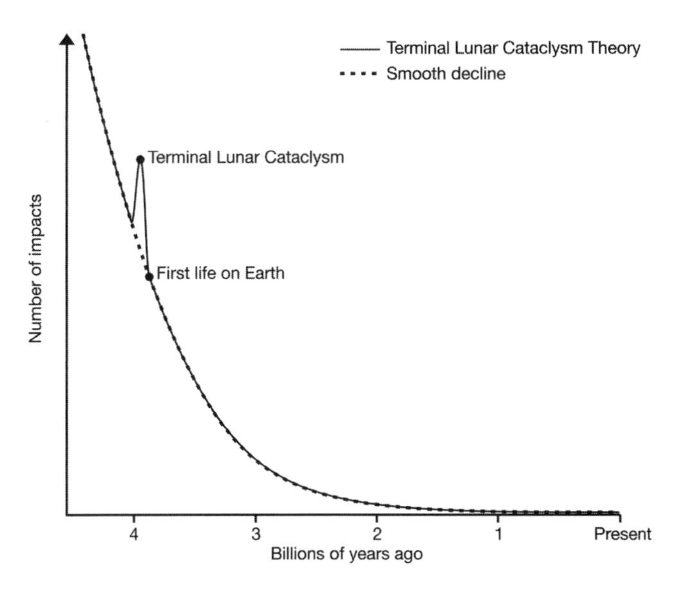

Figure 9.4 The rate of impacts on the Moon is much, much lower today than the earliest times in solar system history. It has long been a matter of debate whether the rate declined steadily or had ups and downs. Radiometric dating of basins on the Moon suggests a spike in the cratering rate near 3.9 billion years ago, dubbed the "Terminal Lunar Cataclysm." While still controversial, evidence for such a cataclysm has been found throughout the solar system by scientists.

about 100 million years old. It is the most recent relatively large crater that has formed on the lunar surface.

Studies of Copernican craters have been of great interest to those scientists who want to know more about the recent cratering flux in the Earth–Moon system. Will the Earth be impacted again any time soon? The answer to that question may lie in the lunar cratering data. Scientists are interested in finding the ages of as many recent craters as possible to see if there have been any changes in the recent flux of impactors. Does the cratering rate continue to fall off uniformly, or are their times of sudden increases in the impact rate? If so, how long do those increases last? There are several large-scale extinctions of species on the Earth, and some speculate that they may have been caused by a burst of impact crater formation. It behooves us to study the Moon's cratering record closely to find out as much as we can.

Lunar Regolith

The surface of the Moon is covered by a layer of broken up bits of rock and glass called the **regolith**. Also referred to as lunar "soil," the regolith is formed by impact processes. It can be as deep as five meters in mare regions, and much deeper, 15 to 20 m deep in the older highlands. This difference in thickness is to be expected given that the highlands have been exposed to impacts longer than the younger mare, and so have had the

Gold Dust and Lunar Volcanoes

Before spacecraft visits to the Moon, the nature of the lunar surface was a mystery. The existence of craters was not in dispute, but the origin of those craters was a matter of fierce debate. For one group of scientists, the craters were obvious evidence of widespread volcanic activity, suggesting that the Moon had once been hot enough to support volcanism. The other group interpreted the craters as of impact origin, seeing no sign of anything other than a permanently dead world occasionally colliding with asteroids and comets.

One of the latter group of scientists was Tommy Gold. Gold not only supported the impact origin for craters, he considered the natural outcome of such an origin: vast amounts of dust. Gold argued that the impacts necessary to make the lunar craters would generate meters or more of fine, loose, unconsolidated dust, enough to swallow probes, spacecraft, and astronauts. While most scientists (even the impact supporters) did not agree with Gold, they could not disprove his theory, either. As a result, early missions to the Moon had to test and confirm that they would not sink into what was jokingly called "Gold Dust." While Gold was wrong about the cohesiveness and pervasiveness of lunar dust, he was proven right that a deep layer of regolith existed on the lunar surface. As for the impact versus volcano theories, in the end both were proven both right and wrong: Most craters are due to impacts, but the Moon clearly had volcanic activity and some volcanic craters do exist on the Moon.

opportunity to be **comminuted** (broken up and mixed) to a greater depth. Large chunks or expanses of solid rock are still exposed on the Moon, but only as boulders thrown from impacts, or along the insides of crater walls or lava channels. The constant mixing, breaking, and turning over of the lunar regolith is called **gardening**.

The lunar regolith has a complex composition due to the processes that create and constantly rework it. Bits of rock from highlands and mare are mixed in with glassy volcanic spherules. Micrometeorite impacts are energetic enough that they form tiny splashes of impact melt, which solidify into impact melt glass and glue other bits of rock and glass together into particles called *agglutinates*. All of these are then available for further reworking, shattering, and melting.

MAGNETIC FIELD

The Moon does not possess a global dipolar magnetic field, such as possessed by the Earth. The crustal rocks do have a magnetic signature that varies in strength across the surface. This magnetization might be residual, acquired by the rocks early in the Moon's history, assuming that despite the small size of the lunar core it could at one time have supported a dynamo. It is also possible that the local magnetization was induced entirely or partially by large impact events.

The lunar magnetic field may be responsible for odd "swirl" features on the Moon's surface. The bright **lunar swirls** all appear to be associated with relatively strong local magnetic fields. The swirls could be areas that are protected from the solar wind, and therefore the upper layer of regolith remains apparently younger, and brighter, longer. Some of the swirls are on opposite sides of the Moon from impact craters, so perhaps impact processes play a part in their formation. However, the most well known swirl, Reiner Gamma, is not directly opposite a large impact crater. There is no one theory that satisfactorily explains how these features form and persist, and so they remain highly enigmatic.

THE ORIGIN OF THE MOON

Neither Venus nor Mercury has any natural satellites. Mars has two Moons, Phobos and Deimos, but they are small and rather like asteroids. With Mars being so close to the asteroid belt, the presence of two possibly captured asteroids around Mars has not been seen as particularly surprising. Even with only four planets in the inner solar system for us to compare to one another, the Earth's Moon is a standout. Why does the Earth have this large, complex planetary body for a satellite when none of the other inner planets do?

This question perplexed scientists for quite some time, with no one scientific model explaining all the data to anyone's satisfaction. The data include the fact that the Earth-Moon system has a high orbital angular momentum. Also, rocks returned from the Moon by the *Apollo* program showed that the Moon was depleted in both refractory materials like iron (materials with high melting temperatures), and depleted in volatiles (materials with very low melting temperatures). Instead the Moon is composed almost entirely of intermediate materials that bear some resemblance to the Earth's mantle. The same rocks also show that the Earth and Moon share a common oxygen isotopic signature.

Before the mid-1970s, scientists were generally considering three possible ideas for the origin of the Moon. The first was that the Moon formed around the Earth, with both bodies simultaneously accreting from the same material orbiting around the Sun. This **co-accretion hypothesis** could explain why the Earth and Moon would have similar oxygen–isotope ratios, but not why the Moon is so depleted in iron. A second idea was that the Moon was actually formed elsewhere in the solar system and eventually captured by the Earth. The **capture hypothesis** could explain the lack of iron, but then was unable to explain why the Earth and Moon would share an oxygen isotopic signature. A last idea, the **fission hypothesis**, was that the early Earth was rotating so rapidly that it spun off a chunk that formed into the Moon. This would explain both the isotope ratios and lack of iron, but not the total angular momentum of the system.

In the mid-1970s researchers began to take a closer look at the idea of a massive planetary collision as the possible origin of the Moon. At a scientific conference in 1984, the **Giant Impact hypothesis** gained the general favor of the planetary science community as the most likely idea of those currently being considered for how the Moon formed. Computer simulations of the time suggested that the last stages of planetary formation around a star would include a host of protoplanets, all running into one another and leaving only a small number of which to survive as the final planets in a solar system. The giant impact that formed the Moon could therefore be envisioned as one of these last stages of planetary formation, where a large proto-Earth was impacted by another large protoplanetary body (sometimes referred to as a "Mars-sized" impactor). Subsequent computer simulations have shown how such an impact would preferentially leave most of the iron from both protoplanets in the newly forming Earth, and throw mantle-like material out to accrete into the Moon.

This hypothesis remains the best one that scientists have so far. Although it is a great improvement over the other ideas, it is not perfect, either. It does not completely explain the depletion of volatile elements, nor the enhancement of some materials like CaO compared to the Earth's mantle. If the giant impact hypothesis does fall out of favor, we currently do not have another, better idea to replace it, and the origin of the

Moon will continue to baffle scientists. But for the moment, it seems that a giant impact is the most likely suspect in the formation of the Earth's Moon.

CONCLUSION

The Earth's nearest neighbor is also one of the best-understood bodies in the solar system thanks to centuries of study by telescope, spacecraft, and visiting earthlings who brought back samples. It appears to be relatively simple geologically, with a plagioclase-rich crust in which giant impact basins have been filled with basaltic lava. Nevertheless, it still has many puzzles yet to be solved, including the details of its origin and impact history. As we mark the 50th anniversary of *Luna 2*'s visit to the Moon and the 40th anniversary of *Apollo 11*'s visit and return, future research holds the promise of delivering those secrets as well as perhaps others yet unimagined.

FOR MORE INFORMATION

To a Rocky Moon, by Don Wilhelms, is the history of the early lunar program as told from a geologist's point of view: http://www.lpi.usra.edu/publications/books/rockyMoon/.

Information about the *Apollo* program, as presented at Goddard Space Flight Center: http://nssdc.gsfc.nasa.gov/planetary/lunar/apollo.html.

Three recent lunar missions: India's *Chandrayaan* 1 (http://solarsystem.nasa.gov/missions/profilecfm?mcode=chandrayaanchandrayaan-1/); Japan's *Kaguya* (http://www.kaguya.jaxa.jp/index_e.htm); and NASA's *Lunar Reconnaissance Orbiter* (LRO) (http://lunar.gsfc.nasa.gov/).

Jay Melosh, a geophysicist studying the giant impact hypothesis for the lunar origin has simulation results and descriptions available online: (http://www.lpl.arizona.edu/outreach/origin/).

10

Mercury: Chemical Connections

INTRODUCTION

At first glance Mercury resembles the Moon; intense cratering and volcanism are evident on both worlds and neither possesses a substantial atmosphere. One might expect them to have similar chemistries, histories, and evolution. But Mercury is surprisingly complex, and differs from the Moon in critical ways. It may hold important chemical secrets to the formation of the inner solar system. Mercury has a weak but global dipolar magnetic field, not unlike the Earth's, implying that Mercury has a partially molten core. Additionally, the bulk density of the planet is very high, which suggests that the core is relatively large. Scientists speculate that Mercury possesses an abundance of certain volatile elements in its core, like sulfur, that have kept it from solidifying. Understanding why Mercury has such unusual chemistry is one of the reasons planetary scientists find it so intriguing. This small planet is an important piece of the puzzle when it comes to understanding how the solar system formed, and what is responsible for the differences between the rocky worlds nearest our Sun.

HISTORICAL BACKGROUND AND EXPLORATION

Mercury is one of the five "classical" planets, known to ancient cultures. Its name in English (and in the Western World) is inspired by its rapid orbit period (88 days), as Mercury in Greek mythology was the messenger of the gods and was particularly fleet of foot. Mercury's orbital period is a consequence of its proximity to the Sun, roughly one-third of the distance between the Sun and Earth. As a result, it is never very far from the Sun as seen from the Earth.

The Planet Vulcan?

We are now confident that Mercury is the largest object of any appreciable size close to the Sun. However, astronomers in the nineteenth century suspected that there might be another planet interior to Mercury. Fresh off of the successful prediction of Neptune, scientists noticed the position of Mercury was slightly different from what was expected, and its orbit was evolving in an unexpected way. Naturally, they interpreted this as evidence of another planet and began the work of calculating its orbit.

As this work was done, predictions of the new planet's positions were made. Some astronomers claimed to have observed transits of the new planet, which was then fed in to make newer predictions. Astronomers were so confident that a planet would be found that a consensus name was effectively decided upon: Vulcan, blacksmith of the gods.

However, observations of Vulcan were never confirmed. New predictions did not help matters. In the early 1900s, Albert Einstein formulated the theory of general relativity, which explained the evolution of Mercury's orbit and removed the rationale for Vulcan. While the planet Vulcan is known not to exist, its legacy remains in the name of the vulcanoids, putative small bodies interior to Mercury's orbit.

It is also rather small compared to other planets, with a mean radius of 2,440 km. The combination of its size and distance from the sun has made Mercury an exceptionally challenging target for astronomers. It has been the most difficult object to observe of all the bodies considered in this volume. These difficulties frustrated attempts to determine Mercury's rotation period. Originally thought to be coupled to its orbital period (just as the Moon's rotation and orbital period are locked together), radar observations in 1965 showed that Mercury's rotation period is actually 59 days long, exactly two-thirds of its orbital period.

The first spacecraft to visit Mercury was *Mariner 10*, launched by the United States. It was the first mission to employ gravity assists, using a close pass to Venus to alter its orbit. Such maneuvers are now commonplace. *Mariner 10* made three flybys of Mercury in the 1974–1975 timeframe, observing roughly half of its surface. The other half remained unobserved until the current century, when the American *MESSENGER* spacecraft had its first Mercury encounters. It is scheduled to go into orbit around Mercury in 2011, spending an Earth year in orbit. Future possible missions to Mercury include the *Bepi-Colombo* mission proposed by the European Space Agency.

GEOLOGIC HISTORY

Mercury's general geologic history is not well understood. We have no rock samples from the planet, as we do from the Moon (*Apollo* and *Luna* programs as well as meteorites) and Mars (meteorites), and until recently we had no global mapping as we do with Venus (*Magellan* Mission). Only in the last few years has the combination of data from *Mariner 10* and the *MESSENGER* mission provided us with imagery of Mercury's entire surface.

Nevertheless, scientists have been able to make some headway in describing the history of Mercury's surface, dividing it into five periods. The Mercurian timeline of five geologic epochs is delineated by crater count data, and they are based only on relative ages, using counts from the Moon as their benchmark. The periods use four specific craters as time markers, discussed below. The relationships between those four craters and other features on Mercury, like lava flows, faults, and other craters, determine in which time period those features are placed. As more in-depth studies of the planet continue, along with better maps and more representative crater counts, the geologic timeline could move, and possibly be redefined. As it is now, however, the order in which the major terrain types on Mercury were formed seems relatively well established. Those major terrain types, mentioned below, have been given names that are descriptive of their appearance, with the exception of the **intercrater plains**, named for their location.

The earliest epoch in the history of Mercury is the pre-Tolstojan followed by the Tolstojan (after the 550-km multiring basin of that name). The first certain event that occurred on the planet was the formation of its crust, at the very beginning of the pre-Tolstojan. There was intense bombardment and basin creation during this period. The intercrater plains also formed during this time, about four billion years ago. They appear to be the oldest terrain on the planet, existing between and around large craters, and predating the **heavily cratered terrain**. These plains probably formed by early widespread volcanism and wiped out some of the very earliest cratering history. About 3.9–3.8 billion years ago, the Calorian era began, and this is about the time the **smooth plains** formed. The epoch was named for the Caloris Basin whose formation and deposits are a major stratigraphic feature of the planet, and it extends to about 3.5 to 3.0 billion years ago. Smooth plains can be found on the floor of Caloris, in a ring around the crater and then in various areas around the planet. The global system of thrust faults, also called **lobate scarps**, appears to have formed after the smooth plains, but it is not certain how long afterward, or even if some may have been forming as some smooth plains were emplaced. These scarps are seen on top of Caloris deposits but not across the **hilly and lineated terrain** associated with that basin. The final epochs in the history of Mercury are the Mansurian and Kuiperian. The Mansurian era is the home of those later impact craters which do not possess bright rays. The Kuiperian era, typified by the bright-rayed crater Kuiper, is expected to have begun approximately a billion years ago and continues to the present day. Since that time, no major events have occurred. Like the Moon, Mercury has seen declining impact rates and the constant production and reworking of regolith.

PLANETARY INTERIOR

Mercury's density is 5.4 g/cc when the effects of self-compression are removed. For comparison, the Earth's uncompressed density is 4.2 g/cc. Knowing the densities of rock and metal (roughly 3.0 and 8.0 g/cc, respectively). we can calculate the fraction of a planetary interior that is taken up by its core. For the Earth, more than half of its radius is taken up by the core, roughly the same as Venus.

The density of Mercury is consistent with a huge core, taking up at least three-fourths of the radius of the planet, making the planet overall exceedingly rich in metals. There are several possible explanations for this chemical difference between Mercury and the rest of the planets. Conditions in the early solar nebula may either have allowed for some sort of mechanical

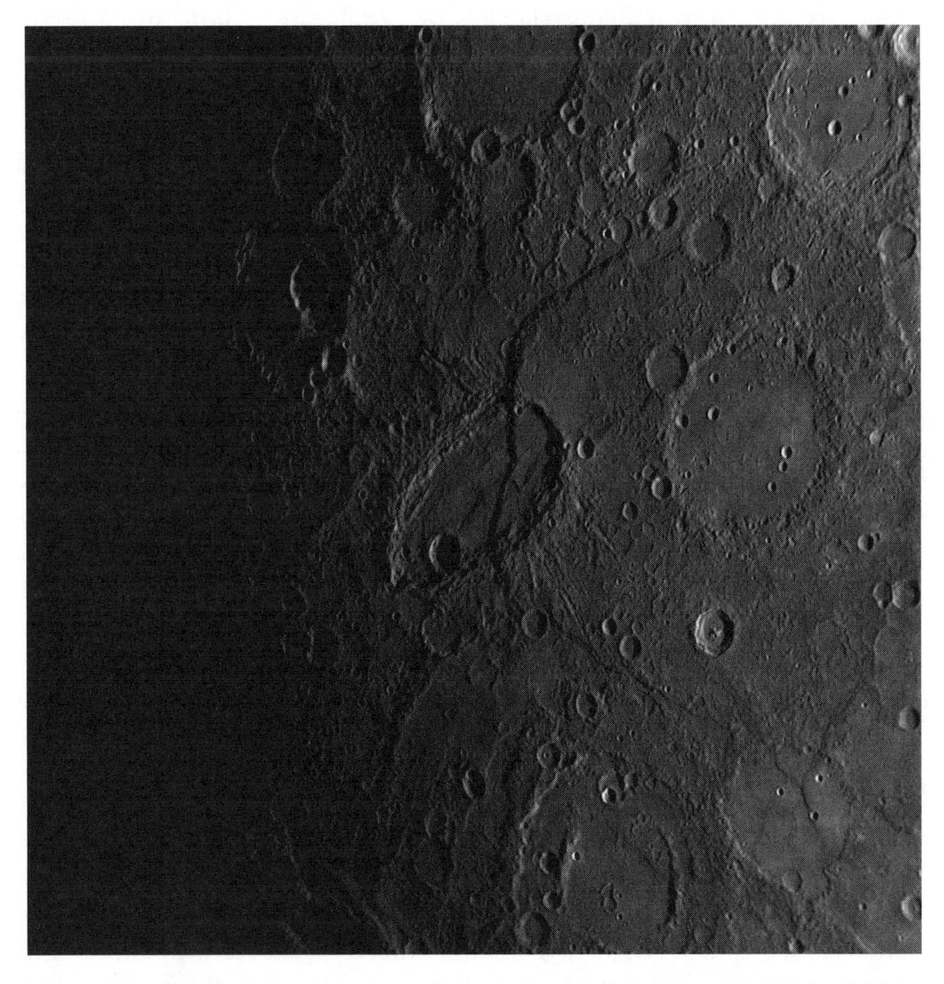

Figure 10.1 Thrust faults, like this one seen by the *MESSENGER* spacecraft, are found across Mercury. It is thought that in part they are caused by the cooling and shrinking of the planet. NASA/Johns Hopkins University Applied Physics Laboratory/Carnegie Institution of Washington.

sorting caused by drag, or for sorting caused by vaporization of volatile elements from the source region or from the protoplanet itself. It is also possible that a giant impact event may have stripped off the top of the mantle and crust, leaving Mercury depleted in these lighter elements. Determining which of these theories is the most likely has far reaching implications for our understanding of the early conditions in the solar system.

While we know crustal formation was one of the first events to occur on Mercury, scientists do not have the information necessary to distinguish between many possible scenarios. It is currently thought that the terrestrial planets were built from pieces with broadly similar compositions. Therefore, absent processes that changed the overall composition of the inner planets, it is thought they should still have largely the same compositions today.

Because of this, it was originally expected that the lunar and Mercurian crusts would be compositionally similar. However, there are observed differences as well. Signatures of Na and K emissions on Mercury might be tied to recent impact events, implying those elements are present in the target material. But there is no strong evidence for a feldspar-like crust, unlike the Moon. On Mercury there is no clear albedo difference between the highlands areas and the plains, as there is between the highlands and the mare on the Moon. There seems to be little difference in the iron content (FeO) from the plains to the rest of the planet, and indeed there seems to be little iron in the surface rocks at all, based on remote sensing. This supports the idea that Mercury is highly reduced (little oxygen bound into the rocks) and that most of the iron is to be found in the planet's large core.

Mercury may originally have been largely molten, or had a magma ocean as is suspected for the Moon. But if so, then there must have been something to change the crustal chemistry, so that Mercury became the body we see today. Giant impacts have been invoked as a possible way to strip Mercury's crust and leave it compositionally different from the other inner planets, but this remains an unconfirmed hypothesis. If there was no magma ocean, then the crust may instead have formed in a more Mars-like constant series of global volcanic eruptions.

In addition to having a large core, part of the core is probably molten, as well. Mercury has a global magnetic field, and a partially molten core is consistent with an internal dynamo in action, as is thought necessary for magnetic field generation. Radar observations of Mercury and tiny motions over the course of decades have independently confirmed the presence of a molten core. However, the size of the liquid region as compared to the solid area is not constrained. Nor is it entirely clear why such a very small planet should still have an appreciable liquid core zone. The most likely explanation seems to be that there are certain volatile elements, such as sulfur, that are present in the core and depressing the melting temperature. This is something of a conundrum, since the amount of volatiles at Mercury should be low, given models for the condensation of the solar nebula and the formation of the planets. Perhaps volatiles were delivered by

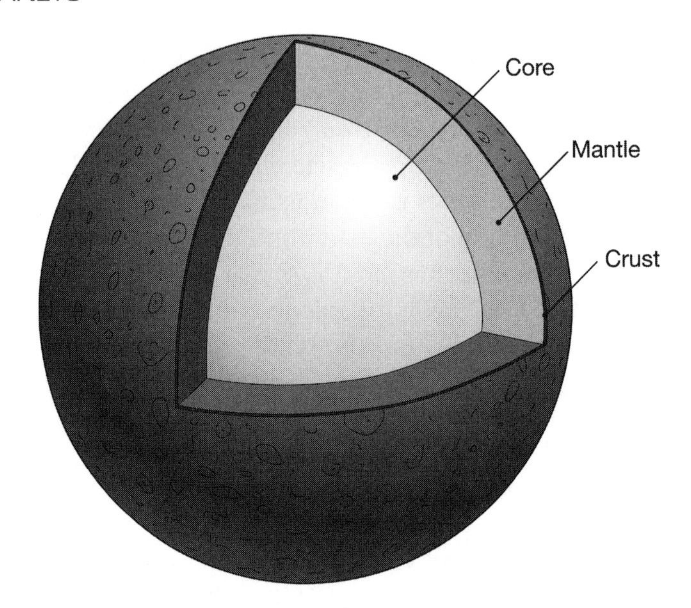

Figure 10.2 Compared to the other inner planets, Mercury has a very large iron core, and a substantially smaller crust and mantle.

planetesimals and comets, or perhaps Mercury formed elsewhere in the nebula. Another possibility is that we are not accurately modeling the nature of the condensation of the nebula and that of early planet formation. Understanding why Mercury has this unusual chemistry has far-reaching implications for how all terrestrial planets form, from crust to core.

ATMOSPHERE

In Chapter 5, we looked at atmospheres in general, and how planets can lose some or all of their atmospheres via Jeans Escape. Above a certain height, and below a certain pressure, atmospheric molecules can have average speeds that are greater than the escape speed, and they are more likely to escape than remain. This height is called the exobase, and the atmosphere above it is called the exosphere. For Mercury (and the Moon), the exobase is effectively at the surface. Atmospheric molecules are more likely to escape than to remain, and if they do remain, they act as particles moving ballistically in Mercury's gravity, whose next action is typically hitting the ground.

Mercury has an exceedingly tenuous "atmosphere" composed largely of hydrogen and helium, as well as oxygen, sodium, potassium, calcium, the recently detected magnesium, and even water vapor. It isn't so much a true atmosphere as it is an exosphere of particles bouncing their way around the planet, into each other, and then into space. Such a process cannot continue without having the particles replenished somehow. There are a few possible sources for the atoms and molecules in Mercury's atmosphere. First, there may be micrometeorites, which vaporize on impact leaving their component atoms to bounce around. Second, ions from the solar wind can stick to the regolith in colder regions, only to be liberated when temperatures increase. Finally, atoms in the Mercurian rocks and soil can be knocked free, or sputtered, and begin bouncing around the planet. The total amount of molecules in Mercury's atmosphere is exceedingly small, fully 15 orders of magnitude less than what is seen on the Earth.

SURFACE FEATURES

The surfaces of the Moon and Mercury have much in common. Each is covered with craters of all sizes, and both possess wide plains of volcanic flood basalts. However, there are critical differences, and these point to the variation in their chemistry, formation and evolution, structure, and surface properties.

Tectonic Activity

The most striking form of tectonic activity recorded on the surface of Mercury is a global distribution of thrust faults. These thrust faults (also called lobate faults or **lobate scarps**) can be several hundred kilometers in length and more than a kilometer high. They are found all over the planet and across all kinds of terrain. These thrust faults are hypothesized to be the result of the entire planet cooling and contracting. But there are other possible models for their formation that could be responsible, or at least partially responsible. The global cooling model meets a problem in that these lobate scarps can only account for a decrease in radius for the planet of about one to two kilometers. However, a cooling event of the magnitude expected would probably have generated a more dramatic decrease in radius. In addition to the lobate scarps, there are other compressional features such as wrinkle ridges and another rarer feature called high-relief ridges.

As for tectonic features due to extensional stresses, rather than only compressional, Mercury has few. The formation and modification of the Caloris basin seems to have produced the only such features on the planet. Within the smooth plains of the Caloris basin, wrinkle ridges are crosscut by extensional graben that may have resulted from uplift of the basin. And there are graben in the terrain exactly on the other side of the planet (antipodal) from Caloris. These features are found in the unusual **hilly and lineated terrain** (also called hilly and furrowed terrain) there, which seems to be a consequence of seismic waves from the massive impact that formed the Caloris basin being focused around the planet and concentrating stresses at the antipodal point.

Analysis of one unique feature, Pantheon Fossae, is ongoing. Nicknamed "the spider," this feature was discovered by the *MESSENGER* team and consists of a number of graben radiating from a central crater. It is not known whether the crater is related to the formation of the graben or coincidentally occurred at the site. While initial theories about its formation relate to stresses from the intrusion of magma, the final word on Pantheon Fossae's formation will likely not be given until well after *MESSENGER* obtains more data.

Volcanic Structures

Mercury has two very important and widespread volcanic terrains, the intercrater plains, and the smooth plains. And yet, finding volcanic structures on the planet has been surprisingly difficult. There are no large volcanic constructs on Mercury, such as the volcanoes on Mars. Given the extent of the flood basalts, it is likely that they erupted from various vents or fissures all over the planet, but those fissures mostly appear to have been covered by the same lavas that erupted from them. Lower-viscosity lavas would not have been likely to build large structures. Studying the volcanic plains of Mercury can be more challenging than for the Moon, because there are very few differences in albedo resulting from different compositions.

The *MESSENGER* flybys of Mercury, however, finally succeeded in locating unambiguous volcanic features. These features are inside the rim of the Caloris Basin, in regions unseen by *Mariner 10*, and the largest is over 100 km in diameter and surrounded by what appears to be bright volcanic ash. Additional data from *MESSENGER* instruments looking at subtle compositional differences between different units, and among the types of material covering the floor of Caloris Basin, confirm the widespread, important nature of volcanism on Mercury as suggested by the *Mariner* data.

Cratering Record

Not surprisingly, Mercurian craters possess all of the general elements of lunar craters; morphologically grading from small bowl-shaped craters to complex craters with central peaks and scalloped rims, and up to the huge basins. Mercury has more than a dozen basins mapped from *Mariner*, and more are expected to be identified as global mapping of the planet progresses. The largest basin on the planet, one of the biggest in the solar system, is Caloris at approximately 1,300 km in diameter. However, there are some differences between craters on the Moon and Mercury. Because of the higher gravity on Mercury, the extent of the continuous ejecta blankets around craters is reduced, as are the distances of secondary impacts from the primary. Fresh craters can have either bright or dark ejecta, depending upon the layers that are impacted.

Comparing the cratering size frequency distributions from heavily cratered terrain on the Moon and Mercury shows that Mercury is relatively depleted in craters less than 50 km. The usual interpretation of this data is that Mercury has wiped out a portion of craters this size in the formation of its intercrater plains.

There are dynamical reasons that the cratering record on Mercury could differ from that of the Moon. A potential factor, if a still theoretical one, would be the existence of "vulcanoids." Vulcanoids are members of an asteroid belt which could theoretically exist within Mercury's orbit, and which could provide an additional population that could impact Mercury but not the other planets. Searches for vulcanoids from the ground and spacecraft have thus far found nothing.

MAGNETIC FIELD

The Moon, Venus, and Mars do not have global dipolar magnetic fields, so it was a surprise to find a weak but Earth-like dipolar field at Mercury. As with the Earth, the field is always changing and is very dynamic, responding to the solar wind as it sweeps by. Mercury gives us our only chance, other than Earth, to study the details that drive a rocky planet's dynamo and power its magnetic field. Ongoing and upcoming missions will provide additional critical information. Magnetic fields are discussed more generally and in more detail in Chapter 8.

BRIGHT POLAR PATCHES

Studies of radar images of the polar regions of Mercury have noted unusual bright patches. These areas all appear inside of impact craters. What this highly reflective material might be is unclear, although water ice is a possible candidate, and many scientists suspect it is by far the most likely one. The craters in which the material has been found could be providing permanently shadowed regions to shield the material from sunlight. Without an atmosphere to buffer the temperature over the planet, any place with direct sun on Mercury will be burning hot, and any place without it will be freezing cold. So a region in permanent shadow could possibly shelter water ice. Interestingly, this phenomenon is also seen on the Moon: radar-bright areas near the poles, with the additional precision available for the Moon showing that the lunar areas are restricted to permanently shadowed regions. In addition, there has been some spacecraft data showing the lunar radar-bright areas also have higher hydrogen concentrations. This is evidence that the bright areas on the Moon are due to water ice, and by analogy is circumstantial evidence for that on Mercury as well.

If it is indeed water ice, there are interesting implications for Mercury's evolution. The water could have come from outside Mercury, for instance being retained from water-rich impactors, with water molecules hopping on the surface (as the atmospheric molecules discussed above) until they reach an area cold enough to stick and remain (called a "cold trap"). Alternately, water could be slowly seeping from the interior of Mercury, and

"Hot Poles"

As noted, Mercury is in a spin-orbit resonance with the Sun, such that during every revolution around the Sun, it completes exactly 1.5 rotations around its axis. Mercury also has a high eccentricity, by far the highest of the inner planets, and a very small axial tilt, much less than one degree (compared to 23 degrees for the Earth).

When these factors were considered together, astronomers realized something curious: at perihelion, the Sun is overhead in only two different places on Mercury. Modelers have found that as a result, these two places experience much greater heating than other regions of the planet, which as a result maintain much higher subsurface temperatures than other areas, as confirmed by radio-wavelength observations. These two areas are naturally, if informally, called the "hot poles."

again hopping until they reach the shadowed regions.

Other than water, the material also might be some type of sulfate, or very cold silicates. More evidence and observations are needed on this phenomenon that will have direct bearing on the nature of the delivery and retention of volatiles on the inner planets.

CONCLUSION

We still know very little about Mercury. Before 2007, less than half of Mercury's surface had even been seen, and *MESSENGER*'s imaging of the rest of Mercury is still being analyzed. At first blush, Mercury appears Moon-like, though it has important differences. Its surprisingly large iron core, magnetic field, and FeO-poor surface composition are not yet fully understood. Mercury's position as the innermost planetary outpost potentially holds great insight into solar system formation, and our knowledge of Mercury can be expected to grow exponentially with *MESSENGER*'s orbital phase.

FOR MORE INFORMATION

Strom, Robert G., and Ann L. Sprague. *Exploring Mercury: The Iron Planet.* New York: Springer, 2003. A discussion of our state of knowledge on the eve of the *MESSENGER* mission.

Head, James W., et al. "The Geology of Mercury: The View Prior to the *MESSENGER* Mission." *Space Science Reviews* 131: 1–4 (2007): 41–84. A more technical view of our state of knowledge of Mercury.

The *MESSENGER* Web site is at http://messenger.jhuapl.edu/index.php, including all of the images they have released as well as the background for their science investigations. *Bepi-Colombo*, the future ESA mission to Mercury, has information at http://sci.esa.int/science-e/www/area/index.cfm?fareaid=30.

Reanalysis of *Mariner 10* images has been undertaken, and results posted and discussed at http://ser.sese.asu.edu/merc.html.

11

Venus: Pressure Cooker

INTRODUCTION

Under the stiflingly thick Venusian atmosphere, 90 times as thick as the Earth's, temperatures exceed the melting point of lead. Amazingly, we have images taken from the surface of this planet because two Soviet landers were able to survive for a couple of hours there before suffering complete failure. Venus is often called Earth's twin, or sister planet, because it is similar to ours in size and orbitally right next door. The original starting materials that formed into the Earth and Venus were not highly different, and yet at some point their evolution diverged dramatically. The Earth became a blue-green jewel with moderate temperatures and abundant water, while Venus became a boiling desert with a blanketing atmosphere of carbon dioxide and clouds of sulfuric acid. Scientists continue to study the reasons for this difference, but the details are still unknown. It is important for us to understand, not just because of general scientific curiosity about the universe, but because we want to ensure our own world remains friendly to life. Understanding the processes that created and maintain the tremendous greenhouse effect on Venus allows us to better model those same processes on the Earth.

HISTORICAL BACKGROUND AND EXPLORATION

Venus is the brightest object in the sky after the Sun and the Moon, and was well known to ancient civilizations. It can remain in the sky for hours after sunset and can appear hours before sunrise, leading to early confusion as to whether the "morning star" and "evening star" were one object or two.

Well after that problem was solved, the fact that Venus goes through phases like the Moon was used by Galileo and others to demonstrate that planets orbited the Sun rather than the Earth.

Other than phases, however, Venus did not show much detail to astronomers who studied it through their telescopes. This led them to conclude Venus was covered in clouds, but unfortunately did not allow much else to be determined with certainty. Some astronomers concluded Venus' rotation period was roughly 24 hours, others that it was tidally locked to the Sun just like the Moon is tidally locked to Earth. Despite the difficulty of obtaining diagnostic data, however, speculation abounded. Some authors, more science fiction than science, considered the thick cloud cover and concluded Venus was likely very rainy and watery. Because it was closer to the Sun, it was almost certainly hotter than Earth. Therefore, they deduced, Venus was most likely hot and rainy like terrestrial jungles.

As the planet that approaches Earth most closely, Venus was the target of many early space probes, and was a continued high priority target through the 1990s. The first big surprise was the discovery that Venus' surface was *much* hotter than expected, with temperatures over 450 Celsius. Rather than a water-rich, humid atmosphere, it was discovered that Venus' atmosphere was largely carbon dioxide, with a liberal dose of sulfuric acid, and a pressure 90 times that of Earth. The Soviet *Venera* program was able to obtain pictures from Venus' surface, showing a barren volcanic plain. The surface was also revealed via increasingly powerful and sophisticated radar observations from Earth and orbit around Venus, with the current best radar map provided by the American *Magellan* spacecraft, which orbited Venus in the mid-1990s.

The fall of the Soviet Union and NASA's focus on Mars has led to a general hiatus in Venus exploration. The European Space Agency's *Venus Express* mission, which arrived in 2006 was the first new Venus mission in over 10 years, though both the *Galileo* and *Cassini* missions had Venus

The *Venera* Program

In the early 1960s, the Soviet Union began a series of missions to explore Venus. They were named "Venera," after the Russian name for the planet. The program provided a great deal of information about Venus, including the only pictures ever sent back from the surface: an achievement recorded by 4 of the missions, *Veneras 9, 10, 13*, and *14*. The images showed flat plains, with platy rocks and some regolith, reminiscent of lava flows. These latter two landers were able to survive for well over their 30 minute design lifetime, with *Venera 13* able to operate for over two hours, by far the longest of any surface probe on Venus.

While the images from the surface were obvious highlights of the *Venera* program, other missions in the series were able to measure the elemental compositions at their landing sites (confirming a basaltic surface), and culminated in a pair of orbiting radars, which provided some of the first detailed views of Venus' geology and the best pre-*Magellan* data available to scientists.

flybys before then and *MESSENGER* has had a Venus flyby since. A community of American scientists interested in Venus have also been organizing their priorities for continued exploration of that planet, and the long-standing Soviet interest in Venus could spark Russian missions there as their planetary program is restarted.

GEOLOGIC HISTORY

The vast majority of the knowledge we have for the surface of Venus was collected via radar or one of the *Venera* landers. The synthesis of these sources can give us very good information about some aspects of Venus, like how rough or smooth it is, but not others, such as any possible compositional variation across its surface. On other inner planets, the relative number of craters in different places is used to determine their relative ages. This tool is not as useful on Venus, however, because the thick atmosphere screens out most small impactors and prevents them from leaving craters, and the number of craters remaining is not sufficient to be able to make simple inferences about relative ages. What can be determined is done by superposition: seeing what features lay atop other features, for instance lava flows erasing a fault or vice versa.

Given that the surface of Venus is relatively young, it does not record all of the events of its geologic history quite so clearly as worlds like the Moon. However, it does appear that the early history of the planet saw the formation of the ancient **tessera** terrain over much of the globe. This terrain was subjected to episodes of compression and then expansion, and then back again. This is a system very different from the Earth's system of plate tectonics movements, with both compression and extension happening at all times. Tessera terrain was also almost immediately subjected to extensive volcanism in the form of flood basalts, creating wide plains. Episodes of compression, tension including tessera formation, wrinkle ridge development and more were sandwiched between large scale lava flows. So, the surface of Venus became one where lava flows would be emplaced over others still showing some tectonic history only to have the new flows also deformed in some fashion. These flows form the regional plains. Subsequent to this was widespread formation of volcanic features visible to the present day along with the sparse global distribution of impact craters.

PLANETARY INTERIOR

The uncompressed bulk density of Venus, added to what we know of the likely surface chemistry and other lines of evidence, indicates that Venus is a differentiated planet with crust, mantle, and core. The *Venera* landers provided compositional data showing that their landing sites were similar

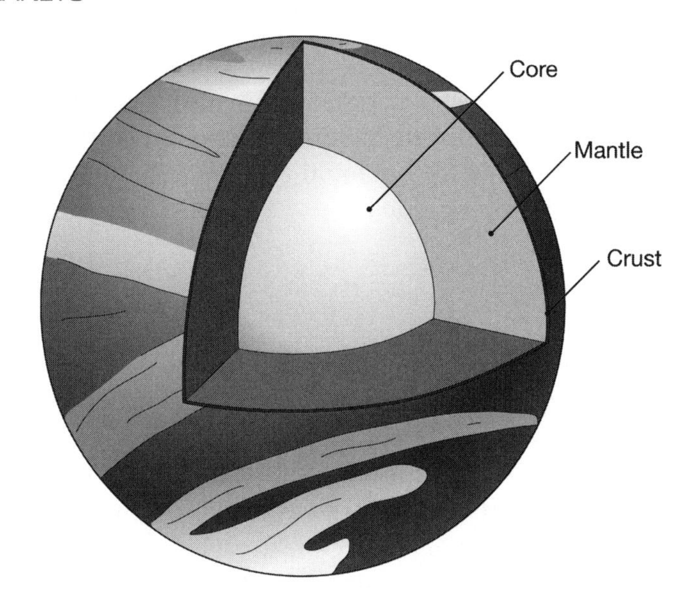

Figure 11.1 Venus, like the other terrestrial planets, is divided into a crust, mantle, and core. It is not known whether the core is entirely liquid, entirely solid, or part solid and part liquid.

to terrestrial basalts, with elemental concentrations suggesting that the rock had melted more than once over Venus' history.

Unlike Earth, Venus does not appear to have its crust split into plates, let alone show evidence of plate tectonics. The lack of plate tectonics on Venus is not due to the planet having solidified or cooled completely, rather, the lack of plate tectonics is an indicator that heat has no efficient way to escape the planet. If the planet had multiple plates in the past, they may have locked up as carbon was removed from the rocks, eliminating a source of lubrication between the plates. Water would also have served some of that function, and indeed some consider this the main reason that Earth *does* have plate tectonics. A dry crust on Venus today is far less mobile than what is seen on Earth. Regional volcanism punching through the crust could continue, but this would not be an efficient means of releasing the interior heat. So we might suspect the planet has the heat to retain its molten core, but not enough pressure to create a solid core within that.

ATMOSPHERE

Venus has a very thick atmosphere composed almost entirely of CO_2 with a small percentage of nitrogen. At the surface level, it is similar in pressure to that of the Earth's oceans one kilometer deep. There are winds on Venus, with speeds that differ greatly depending upon the elevation. At the surface they are slow and ponderous, much less than the typical wind speeds at the Earth's surface. In spite of that they can have some effect on the surface.

The high density winds can force some surface debris to move in front of them, just as how on Earth a slow-moving river can move material more effectively than the same-speed wind. As elevation increases, the wind speed picks up sharply until it reaches 100 m/s at roughly 60–70 km height. These winds are largely east-west rather than north-south, and serve to very effectively distribute heat across Venus, effectively circling the planet every few days. Despite the fact the time from noon to noon is over 110 Earth days, the temperature on Venus is nearly the same across its surface. The difference in the rotational speeds of the atmosphere compared to the surface is called **superrotation**.

Figure 11.2 The clouds of Venus are featureless in visible light, but show much greater contrast in the UV. This image, taken by the *Pioneer Venus Orbiter*, shows the clouds, as well as illustrating the difference in wind speeds with latitude on Venus.

The winds have been directly measured by observing Venus' well-known clouds, which prevent us from directly seeing the surface. Over most wavelengths the cloud tops are bland, but detail can be seen in the ultraviolet wavelengths, where observations have been focused for a few decades. The clouds of Venus are composed of sulfuric acid droplets and sulfur dioxide—much more poisonous than the clouds of Earth. They reflect roughly 75 percent of the light that falls on them, making the surface of Venus quite poorly lit. The cloud deck is mostly restricted to 60–70 km in height, although there is also a haze layer stretching another 10 or so kilometer beneath the clouds.

As mentioned above, the dense atmosphere is an excellent way to filter out smaller impactors seeking the surface. Most blow up in the atmosphere, while others that only just make it to the surface explode, creating a scattershot or blasted pattern on the surface. Some craters are surrounded by

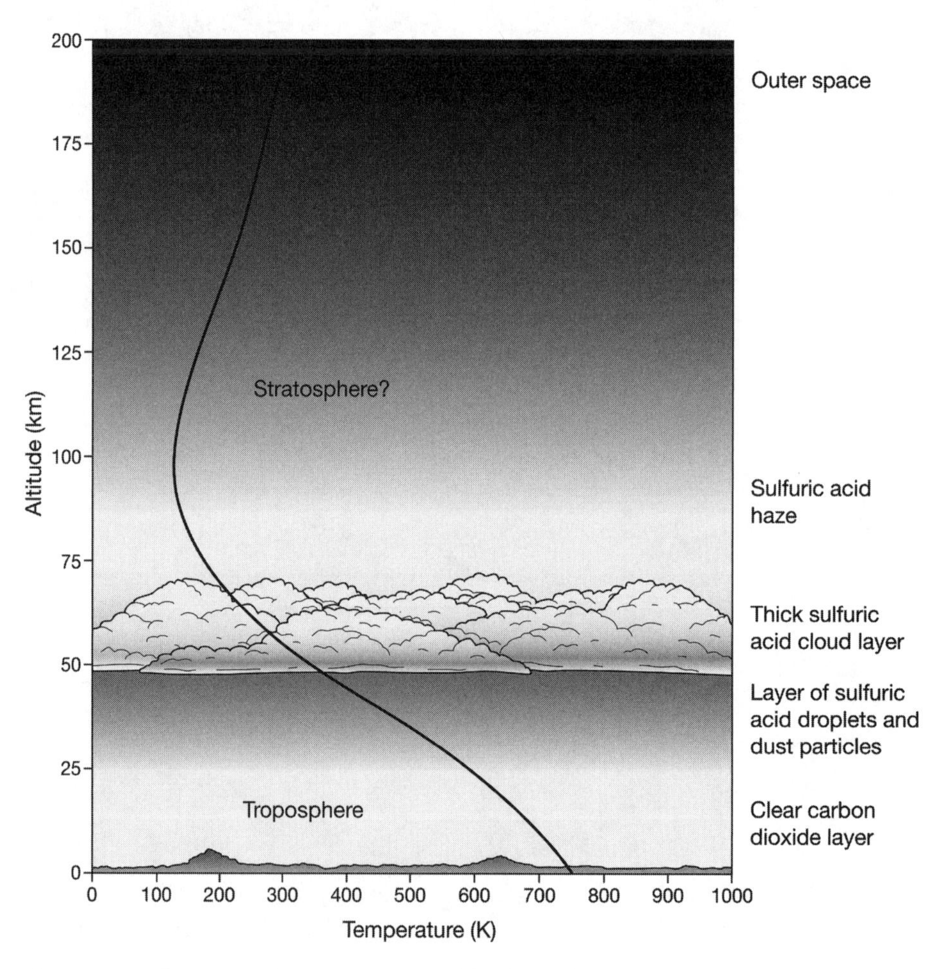

Figure 11.3 The atmosphere of Venus, unlike that of the Earth, only has two layers: a troposphere with temperature decreasing with altitude, and a stratosphere with temperature slowly increasing with altitude. Earth-like temperatures and pressures are reached at roughly 50 km altitude, within sulfuric acid haze and clouds.

parabola shaped features that are possibly ejecta particles swept forward in the wind. This has also been used to estimate the wind speeds near the surface. The signature of lightning has been detected on the planet; this may be more typical atmospheric lightning, or as some suspect, an indication of ongoing volcanic activity.

SURFACE FEATURES

The surface of Venus is dominated by volcanism—more than 80 percent of the surface is blanketed with volcanic plains. We do not know if that has been the case throughout the history of the planet, since Venus has successfully resurfaced itself one way or another in the last 200 to 700 million years. What we can deduce from orbital data shows a world with some impact craters, but mostly volcanic domes, lava flows, almost 170 giant volcanoes, and formations created by internal processes called "coronae." There are a few mountain ranges, but overall the planet is quite topographically flat compared to the Earth. At its most basic, Venus is often referred to having two major land units, the lowland plains (Planitia) and the higher relief regions (Terrae). The Terrae cover roughly 10 percent of Venus' surface, with the plains covering the remainder.

Tectonic Activity

The surface of Venus holds plenty of evidence for tectonic activity. This includes the "continents" of Ishtar Terra and Aphrodite Terra, respectively. This Tessera Terrain is a very old terrain characterized by intersected grooves and ridges, and what now remains of it covers less than a tenth of the planet. Older terrain is covered by lava flows that themselves may have plenty of tectonic features such as fractures and ridges. As a whole these form the regional volcanic plains. Internal processes have also generated

The Names of Venus

The surface features of Venus were entirely unknown until the 1960s. Starting in 1964, scientists at large U.S. radar facilities were able to obtain data of sufficient quality to identify distinct regions on Venus. The group working at Goldstone, in California, simply identified spots by Greek letters. A different group working at Arecibo in Puerto Rico honored scientists who worked with radio and radar with named features.

The prospect of needing thousands of names prompted a review of the nomenclature of the Venusian surface, and it was decided to name them after prominent women of mythology and reality from Gertrude Stein and Cleopatra to Ishtar, Lakshmi, and Aphrodite. However, three names were held over from the initial radar days: Alpha and Beta Regio, and Maxwell Montes (named after James Clerk Maxwell).

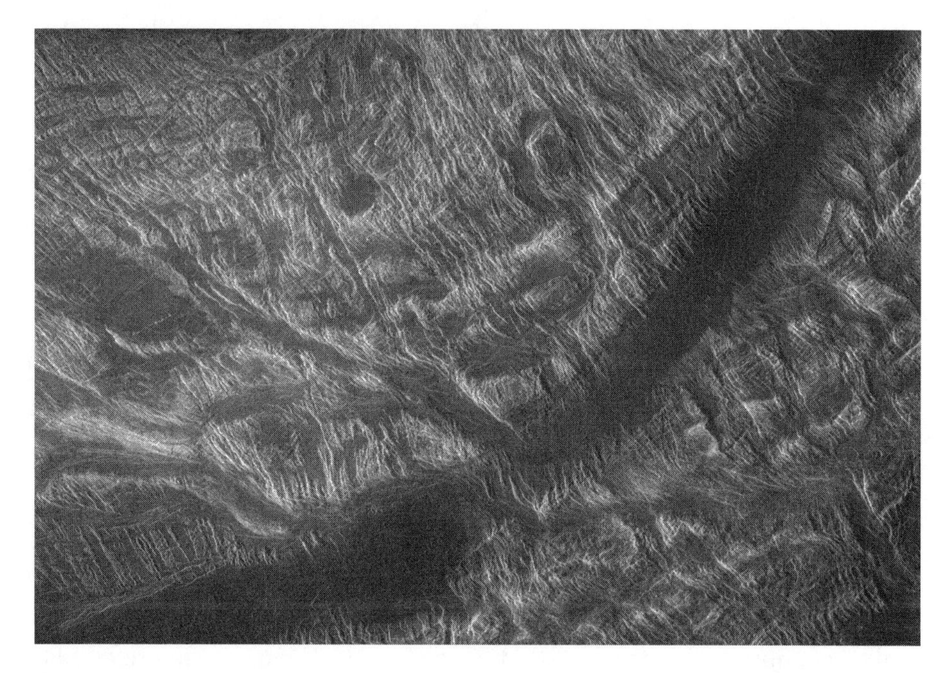

Figure 11.4 Ovda Regio, shown here, is one of the venusian highlands. As with the other highlands regions on Venus, it has been extensively faulted and folded.

the quasi-circular features called coronae (singular: corona). Coronae appear to have formed by upwelling magma causing a large dome, which then deflated in the middle as the magma escaped or drained. Well over 100 coronae have been observed on Venus, typically hundreds of kilometers in diameter, though the largest (Artemis Corona) is roughly 2600 km in diameter. Some scientists interpret Artemis Corona as formed via regional-scale plate tectonics.

Certainly the Venus of today does not possess plate tectonics as does the Earth. Perhaps it did in the past. There is some speculation that Venus did once possess multiple mobile plates, but as the planet cooled, the plates ground to a halt and eventually locked, creating a one-plate planet. If this did occur, then heat from the interior would have found a new barrier as it attempted to escape the planet. This may have been the source of the sudden and intense bout of volcanic melting and extrusion that has covered the surface of the planet, and erased the record of all but the most recent impact craters. But the necessity of requiring just one such resurfacing event is debatable.

Volcanic Structures

The surface of Venus is dominated by volcanism. Using crater counts to estimate ages to the best degree possible, it appears that the bulk of the structures formed about 450 million years ago, plus or minus some

Figure 11.5 The surface of Venus was imaged by the *Venera* landers, showing a barren volcanic landscape. The lander itself is at the bottom of the image.

250 million years. This range of age from 200 to 700 million years comes from issues in calibrating the absolute age of surfaces from the Moon to Venus, in the interpretation of models of how the interior of the planet has been cooling, and in the nature of the volcanic features themselves.

When the global surface of the planet was finally mapped, scientists at first thought that the entirety of Venus had been catastrophically (that is, all at once) volcanically resurfaced about 500 million years ago. Current thinking, however, suggests that a single catastrophic event is not necessary to produce the surface we see today. The planet's surface holds evidence for several episodes of resurfacing, or perhaps continuous resurfacing with decreasing intensity. While it appears that nearly all of this activity ended by 200 million years ago, it is not impossible that there is some small-scale volcanism taking place on Venus even today. These various hypotheses continue to be debated.

What is not debated is that Venus is a volcanic wonderland, with wide volcanic plains, giant volcanoes, volcanic domes, shield volcanoes, caldera, small cones and more. The coronae on Venus appear to be related to both tectonics and volcanism. They are possibly a result of rising magma plumes from under the surface, as they rise they deform the crust, and tectonic cracks and related features develop as a result.

Cratering Record

Venus has impact craters randomly scattered over its surface. Unlike the Moon or Mars, there is not a definite distinction between very heavily cratered, older terrains and lightly cratered, younger terrains. The whole of the planet appears to be at approximately the same age, which is a substantial piece of evidence in favor of massive resurfacing. The oldest surfaces are likely to be no more than 700 million years old (although there may be very small features quite a bit older than a billion years of age), while most appear to cluster around 500 million years. The smallest craters on the surface are about 3 km in diameter, since anything that would create a smaller crater will disintegrate or burn up in Venus' atmosphere. The very largest craters are also missing. Craters remain below a few

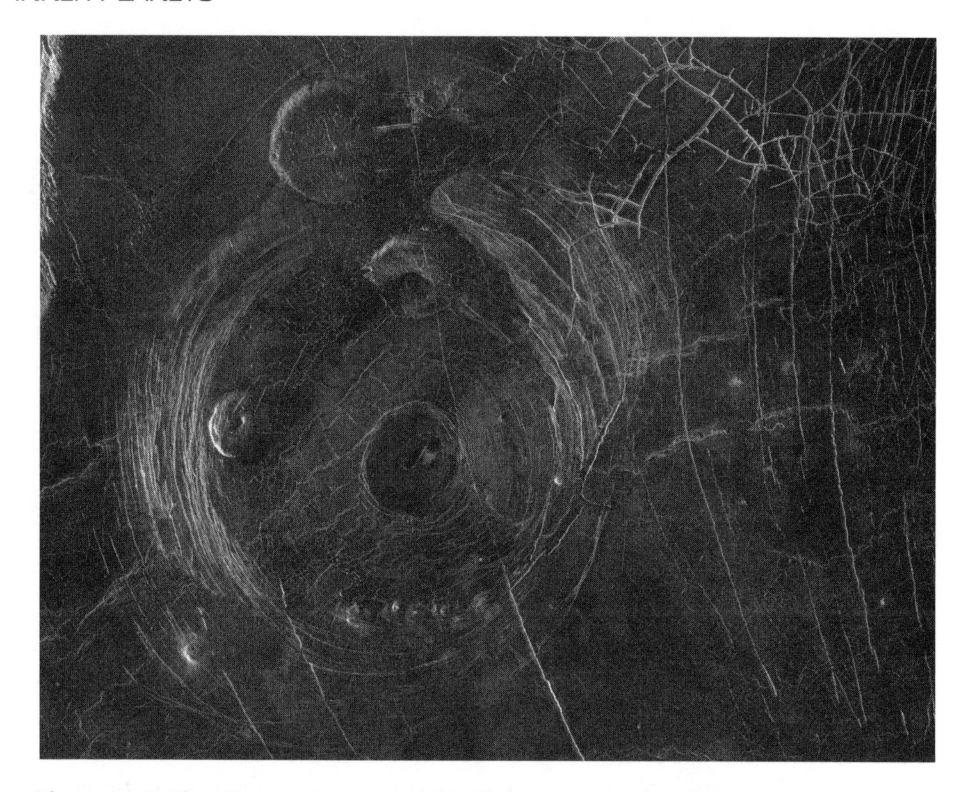

Figure 11.6 Aine Corona is an example of a common volcanic structure on Venus. Coronae are thought to have inflated as magma was fed in from below, then collapsed after the magma drained away leaving a circular depression. Faults crisscross the area.

hundred kilometers in diameter. Basin forming impacts are usually quite old, and their absence suggests that Venus has eliminated its very ancient history.

Crater counting is very useful on planets with a variety of terrains and ages, that have collected clearly different representations of the impactor flux of the time. As mentioned earlier, Venus is not a great place to try to use crater counting to find the ages of small extents of terrain, or of larger extents of nearly similar age. There simply are not enough craters available (only about 1,000 impact craters on the whole surface) to allow statistical analysis of crater counts to make fine distinctions in the ages of the volcanic terrains.

MAGNETIC FIELD

As discussed in the chapter about magnetospheric processes, a large active dynamo can produce a planetary magnetic field. This is exactly the situation on Earth, where our rapidly rotating and partially liquid core powers an impressive global magnetic field. In spite of Venus's large size, it does not have a global magnetic field. There is a form of minor magnetic field introduced by interactions between the highest parts of the Venusian atmosphere and the solar wind, but is not significant compared to the terrestrial or Mercurian fields. There are competing theories why this is the case. The core of Venus could be entirely molten, with no solid center. This could be because the planet may not be quite big enough to produce a solid core the way the Earth does, through massive interior pressure. Another related possibility is that a liquid core might be isothermal—the same temperature all through, in part because of the lack of plate tectonics would efficiently trap Venus's heat and insulate the core. An isothermal core would inhibit any kind of convection taking place, without which a dynamo would not operate. A further factor could be the very slow spin of the planet. Venus is quite odd in that it rotates very slowly—and in the opposite direction of all the other inner planets. This is called retrograde rotation, and what could have caused it is uncertain. Scientists are hesitant to invoke another instance of a giant impact, but it is not impossible that this is the source of the planet's strange rotation.

Remnant magnetism has been seen on other planetary bodies and used to infer the presence of magnetic fields in earlier times. It is conceivable that Venus once had a powerful magnetic field like the Earth's, and that it died away. However, it is unlikely that any rocks on the surface would retain residual magnetism due to the very high surface temperatures, and any such information is either lost to us or awaits the ideas of a clever planetary scientist of the future.

PAST WATER ON VENUS

There is some indication that Venus once had abundant water. Unlike Mars, which preserves ancient channels and river valleys, Venus has no water-cut surface features. So why do scientists think it once had water? First of all, the theories for planet formation predict that the Earth and Venus started out with roughly similar amounts of water. The exact fraction derived from initial accretion and the fraction delivered from comets might vary, but whatever the source, Earth and Venus probably shared it. But there is more compelling evidence, and this evidence lies in the ratio of two different isotopes of hydrogen found in the atmosphere of the planet.

Hydrogen was formed in the Big Bang, and it so happens that while most hydrogen is "regular" hydrogen, with one proton and one electron, about one in every 10,000 hydrogen atoms also happens to have a neutron in the nucleus. This kind of hydrogen (with a neutron added) is called **deuterium**, and its properties differ from ordinary hydrogen essentially in that it is more massive. The ratio of deuterium to hydrogen (the **D/H ratio**) is therefore constant in nature. No matter where you find hydrogen, you will find this same D/H ratio unless some specific process altered the ratio, and there are only a few processes that can manage that.

Hydrogen is not generally incorporated in large quantities in the rocky planets simply in its elemental state. It is too volatile for small planets to hold on to. Instead, hydrogen is found, and very abundantly, as a component of water, methane, and ammonia. It is also found bound to rocks, but water is generally needed already to drive that reaction. Again, for the inner planets, methane and ammonia are not nearly so abundant, and so water vapor becomes the dominant place to find hydrogen in the atmosphere. Since water molecules have two hydrogen atoms, about one in every 5,000 water molecules contains a deuterium atom.

Imagine now that you have a planet, much like the Earth, except for its position nearer to the sun. The planet is warm enough that the amount of water in the atmosphere (as opposed to on the surface) is increased by evaporation and similar processes. Meanwhile, CO_2 is also released by active volcanic processes. Both H_2O and CO_2 are particularly effective greenhouse gasses. The result of the greenhouse effect is to increase the temperature, and as the temperature goes up, more H_2O and CO_2 are released, and the temperatures go up more. This is a greenhouse that has "run away"; with no other processes available to buffer or mitigate, the greenhouse increases without stopping, superheating the surface until all the water is completely gone into the atmosphere, along with the CO_2. CO_2 is a relatively heavy molecule, and isn't easily lost to low gravity. Water is also heavy, but is subject to a process known as **photodissociation.** UV radiation from the sun can break the water and hydrogen apart. The oxygen from the dissociated water would be left to interact with rocks on the surface of the planet, and the hydrogen would be subject to erosion either because the gravity was too low to hold it or because the solar wind was available to strip it away.

This is essentially the history of Venus's water. The CO_2 remains as the overwhelmingly dominant molecule in the planet's atmosphere. The oxygen is bound into rocks, and the hydrogen has been largely lost to space. But it was not completely lost. Some hydrogen remains, preferentially deuterium, since it is the isotope that is less susceptible to erosion because it is heavier. The D/H ratio of the hydrogen in Venus's atmosphere indicates that a vast deal of hydrogen was once on the planet—and as we know, it must have been in the form of water. The Venusian D/H ratio tells us Venus once had plenty of water, and lost it all.

What does this tell scientists? It is possible that a rather small change in Earth-Sun distance for a planet can radically alter the interactions of the systems and processes that control the planetary environment. The greenhouse effect is a good thing; it means that our atmosphere can keep our planet at a nice warm temperature. But if somehow the delicate cycles of carbon and such on this planet were disturbed, could the greenhouse run away? This is one of the reasons scientists are particularly interested in greenhouse gas emissions, and in understanding just how much is too much. Venus could be seen as the ultimate warning of the consequences of climate change.

CONCLUSION

Venus was once thought to be "Earth's twin." Its size, solar distance, and the presence of an atmosphere led scientists to believe that once we could peer beneath the clouds, we would find a planet much like our own. The truth has turned out to be much different than expected: a torpid volcanic landscape and a stagnant, poisonous atmosphere giving way to ferocious winds whipping clouds of acid high above. Yet Venus remains a fascinating place and an object lesson of how small differences in initial conditions can lead to very different outcomes. As it is explored further in coming decades, Venus will no doubt have much to teach us about how the Earth became the oasis it is, and how we might preserve it.

FOR MORE INFORMATION

The USGS has an exhaustive list of Venus-related (and geology-related) technical readings listed at http://astrogeology.usgs.gov/Projects/PlanetaryMapping/VenusMappers/Reading.html.

The USGS also hosts a set of Web pages with downloadable/printable Venus maps and interactive Venus maps, as well as lists of different surface features: http://planetarynames.wr.usgs.gov/jsp/SystemSearch2.jsp?System=Venus.

Don Mitchell, a retired researcher, has put a great deal of time and effort into reanalyzing the pictures taken by the *Venera* missions using computing and image processing techniques unavailable when the data were first taken. This information and data are hosted at http://www.mentallandscape.com/V_Venus.htm and http://www.mentallandscape.com/C_CatalogVenus.htm, which also includes a vast amount of information about the *Venera* program and other Soviet planetary missions.

The *Magellan* mission, though it was finished at the very dawn of the World Wide Web, still has a web presence at http://www2.jpl.nasa.gov/magellan/.

The Venus Exploration Analysis Group acts as a liaison between the scientific community and NASA, and their Web site (http://www.lpi.usra.edu/vexag/) includes much of the scientific community's thinking about the future direction of Venus research and mission planning.

Venus Revealed: A New Look Below the Clouds of Our Mysterious Twin Planet by David Harry Grinspoon (Perseus Publishing, 1997) takes stock of our understanding of Venus after *Magellan* in a popular-level work.

12

Mars: World of Dust and Wind

INTRODUCTION

The red planet we know as Mars was once a very active world. There were massive volcanoes, wide rift valleys, and extensive outflow channels. This was a place with intense volcanism, powerful marsquakes, and rushing water flow. But that is not the Mars of today. Today the water is gone, and the volcanoes are quiet. Instead, almost all of the changes on the planet are driven by interactions between the surface and the atmosphere. Although Mars's atmosphere is thin in comparison with the Earth's, it is more than substantial enough to churn up global dust storms. Aeolian processes dominate, with dust being removed and then redeposited all over the surface. Dunes are commonplace features, as are the tracks from tornado-like twisters known as dust devils.

Even so, understanding the history of the planet, and thus putting its present into context, is no simple endeavor. Mars is a complicated place. Contributing to its interesting geologic history is a chaotic planetary "wobble" that has influenced weather, seasons, and climate on the red planet for its entire history. The Earth has a stable axial tilt of about 23 degrees. It is possible that this tilt has been stable for so long because of gravitational interactions with our large satellite, the Moon. Mars does not have a constant axial tilt. It can vary from small to large, and in a random fashion. Therefore, different parts of the surface of the planet have been heating and cooling over time, and so seasonal effects, along with weather, have varied. In addition, the planet's volcanic history has played a role in episodically heating the crust and releasing water onto the surface and gasses into the atmosphere. The result is that Mars has not gone uniformly from a time of a thick, warm atmosphere with abundant water to a thin, cool atmosphere and subsurface ice. Instead, there appear to have been multiple episodes of ice melt, water flow, and increased atmosphere.

An issue that always arises when studying Mars is the problem of trying to determine what is happening actively today and what is a relic of the past. Craters, volcanoes, and even wind streaks could be merely the remains of a more active history. And some of them, even if on small scales, could be forming and changing today. It can be challenging for scientists to decide which are which with certainty.

HISTORICAL BACKGROUND AND EXPLORATION

Mars, the fourth planet from the Sun, is a distinctive sight in the night sky. Its reddish appearance gained it an association with the Roman god of war, after whom the planet was named. Mars also has a relatively high eccentricity. This fact, combined with Tycho's careful observations of Mars, helped convince Johannes Kepler that Mars's orbit could not be perfectly circular and led to the development of the laws of orbital motion still used today.

In the telescopic era, Mars has been also been a frequent target. The visibility of Martian surface features and seasonal changes made Mars quite popular among astronomers by the nineteenth century. The seasonal changes were often interpreted as vegetation growing and dying throughout the Martian year, and the probability of life on Mars was considered fairly high through the middle of the twentieth century. The search for life on Mars, or its precursors or remains, continues to be a major impetus for current Mars exploration.

Spacecraft visits to Mars have been made since the 1960s. The United States has dominated exploration of Mars, starting with *Mariner 4.* The first missions happened to fly by heavily cratered and relatively bland terrain, leading to great surprise when the *Mariner 9* orbiter found a planet with great geological diversity. This was followed by the *Viking* orbiters and landers, which revolutionized our understanding of Mars. *Viking* was designed to search for life, and it found none. After *Viking,* the NASA Mars program slowed considerably, and was also hampered by several mission failures before reaching a more successful period beginning in the mid-1990s with orbiters, landers, and perhaps most notably, rovers. In addition to the American missions, the Soviet Union had a partially successful mission (*Phobos 2*) after a string of unsuccessful missions, and the European Space Agency's *Mars Express* mission has spent several fruitful years at Mars.

Future Mars exploration will likely include larger, more capable rovers, with scientists hoping ultimately for sample returns from Mars. Mars is also a possible target for human exploration. Advocates of human exploration of Mars argue that it is the next obvious destination after the Moon, and

Martian Canals

Some astronomers, straining to record minute details on the Martian surface during fleeting moments of extremely steady skies, reported a network of lines and arcs connecting darker areas. An Italian astronomer, Giovanni Schiaparelli, interpreted these lines as similar to water-carrying channels on Earth, and applied the Italian name for such features, "canali." When read by English-speaking scientists, *canali* was misinterpreted as meaning "canal" rather than "channel," implying construction by intelligent life. Some imaginative astronomers, most notably Percival Lowell, constructed scenarios connecting a number of Mars observations and imagined a dying civilization transporting water from the poles to thirsty equatorial cities. This concept was enormously influential in shaping the public's view of Mars, with echoes of it appearing in movies to the present day. However, the canals themselves have proven to be optical illusions, never seen by spacecraft or modern observers.

that the time and effort involved in such a mission makes a permanent base the best and most cost-effective way of achieving those goals. Opponents of human exploration of Mars believe that the robotic program alone can reach the science goals, and that the uncertainty and cost makes a human mission to Mars ill-advised.

GEOLOGIC HISTORY

Martian history is split into three major epochs: the Noachian, Hesperian, and Amazonian. The timing for each of these epochs is not absolute. They are defined by crater count data, which must be calibrated using counts from the Moon. Any differences in interpretation of cratering flux will change the absolute ages inferred for the epochs.

The most ancient period, the Noachian, saw the formation of the large impact basins and the creation of the heavily cratered southern terrain. This epoch is best dated from about 4.5 to 3.6 billion years ago, though the younger end is somewhat uncertain. The planet's crustal dichotomy (see the following section on tectonics) probably formed early in this period. The Hesperian followed the Noachian and lasted until approximately 1.8 billion years ago. It was characterized by extensive volcanism, including in the Tharsis region, continued cratering, and the release of massive amounts of water, forming wide outflow channels. The massive canyon region of Valles Marineris began to form in the Hesperian. In the Amazonian, the most recent epoch, volcanism continued in the regions of Olympus Mons and the Elysium plains, although not as extensively. Cratering continued to fall off, and Valles Marineris was modified by additional tectonic and Aeolian processes. Wind began to dominate the surface activity of the planet, leaving behind massive layers of sediment in some areas and causing extensive erosion in others.

PLANETARY INTERIOR

Mars is a chemically differentiated planet, with a separate crust, mantle, and core. The bulk density is unexpectedly low, 3.9 cc (Venus and Earth are at approximately 5.3 cc). This might otherwise imply a relatively small iron core, but models suggest this is not the case. Geochemists have had their modeling efforts greatly aided by the existence of meteorites that originated on Mars (formerly called the **SNC meteorites**, now simply called "Mars meteorites").

Study of the Mars meteorites tells us not only about the surface of Mars, but also about the interior: The specific distribution of elements found in minerals can be compared to the original distribution (as seen in primitive meteorites), and used to deduce the elements in the mantle and core. Given those elements and the pressures and temperatures expected inside Mars, geochemists can model the composition and minerals that should be in Mars's interior and then compare that to more data when it becomes available.

What the geochemists expect is for Mars to have several different compositional layers from the surface down to its center. The crust, comprising the top 10–50 km, can be directly observed from the Earth and from spacecraft, and consists of the products of volcanic eruptions and more recently weathered material. Beneath the crust is a layer of silicate minerals (olivine, garnet, pyroxene) in the mantle. As pressure increases, those minerals change forms and structures but maintain their compositions as spinel and majorite, and finally change to a mix of silicates and oxides. A bit over 2000 km below the surface, the composition changes drastically as an iron core, mixed with other elements, constitutes the remainder of the interior.

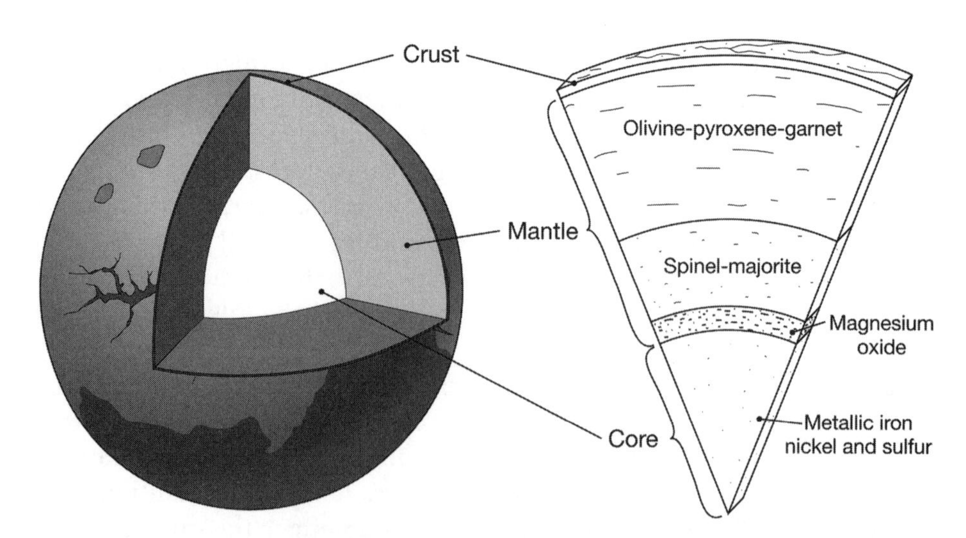

Figure 12.1 The interior of Mars consists of several layers of rock, with their structure changing with depth and increasing pressure. Below 2000 km an iron core is found, with small amounts of other elements.

ATMOSPHERE

The dense early atmosphere of Mars, composed largely of CO_2, is now possibly partially bound up in the Martian rocks. The formation of carbon-rich rocks on the Martian surface could have sequestered atmospheric carbon, but scientists are still looking for evidence of these rocks via remote sensing (some of those minerals have been found in the Mars meteorites). The atmosphere could also have been lost because the planet has low gravity, and because it lacks a magnetic field to shield it from the erosive effects of the solar wind. Some theories suggest that large impact events may serve to strip away vast parcels of atmosphere directly into space. Volcanism may have played a role in the density of the early atmosphere, since volcanoes can be a source of CO_2. After volcanism ended, there would have been no real source to replenish CO_2.

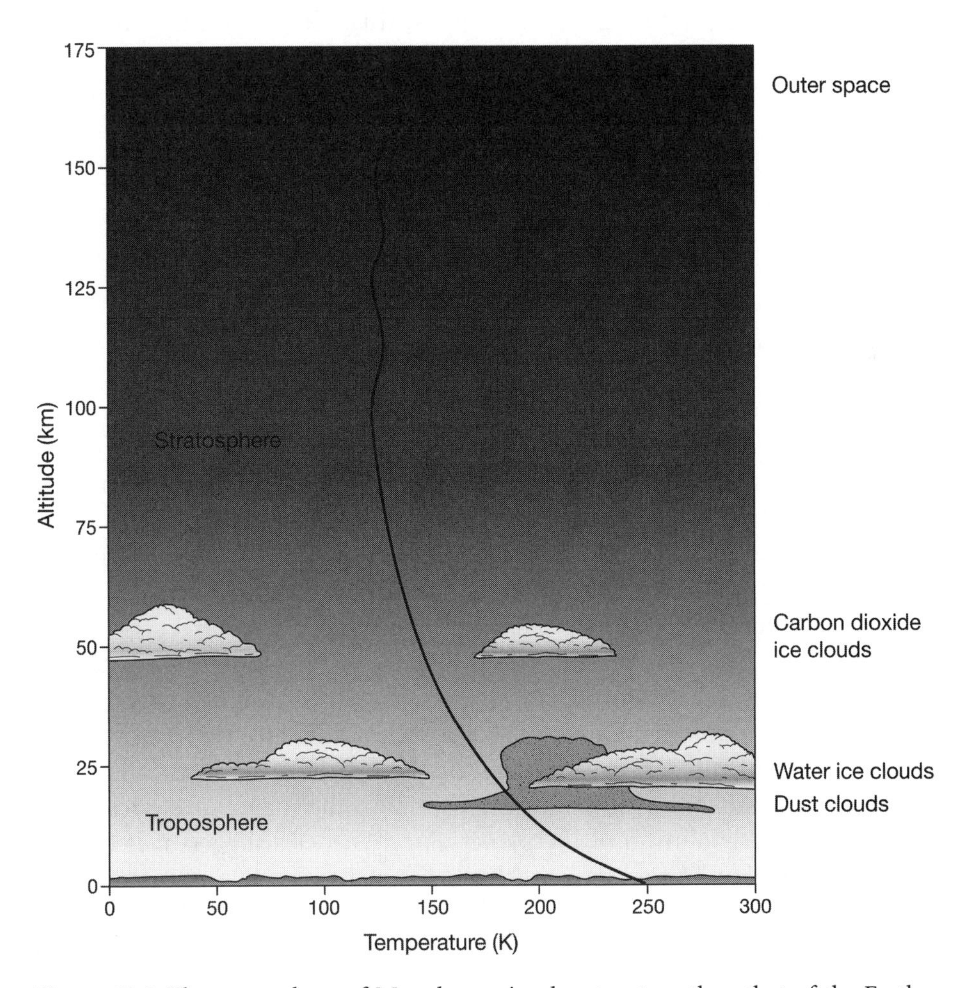

Figure 12.2 The atmosphere of Mars has a simpler structure than that of the Earth, with only two major regions. Dust dominates the near-surface region, while water and carbon dioxide clouds can be found higher in the atmosphere.

The atmosphere of Mars today is one of thin CO_2 and traces of water vapor. It circulates about the globe in Hadley cells and patterns much like the trade winds of Earth. Variations in solar heating during the course of the Martian year cause seasonal changes in those currents, and allow for CO_2 to move between the polar caps and the atmosphere. The winter pole of the planet is so cold that some of the carbon dioxide from the atmosphere freezes right out, and this drops the atmospheric pressure substantially. Mars does not have the same highly stable profile of temperature versus height seen on the Earth. Instead, the vertical profile of the Martian atmosphere can vary depending on the season, the quantity of dust in the atmosphere, and the latitude. Dust is a critical issue when considering the atmosphere of Mars. High dust concentrations usually mean higher temperatures.

Clouds have been seen on Mars, though no precipitation other than dust has been recorded by scientists. However, the transport of dust is of great importance, as just mentioned. Large-scale dust storms can cover practically the entire planet. On small scales, dust devils can be extremely common, and much larger than those found on Earth. Interestingly, the dust cycle has even had an influence on the landers and rovers exploring Mars. That is, the solar panels they use for power become less useful as time goes on and they are covered with dust, but on occasion, the *Mars Exploration Rovers* have encountered dust devils that have cleaned the panels and given them a new lease on life.

SURFACE FEATURES

One of the most striking surface features on Mars is on the scale of the entire planet. One hemisphere, predominantly the southern one, is covered with older, higher, more heavily cratered terrain. In the northern hemisphere, one mostly finds wide, smooth volcanic plains with noticeably fewer craters. The northern hemisphere is one to three kilometers lower than the southern, and it is suspected that the crust is 15 to 20 km thicker in the south. The north is generally referred to as the northern lowlands, or plains, and the south as the southern highlands. In some places, these two provinces are separated by a **scarp** (a steep cliff face) about two kilometers high. This striking difference between one half of the planet compared to the other is referred to as the **Martian crustal dichotomy**, and the cause of it remains one of the major mysteries of the formation and evolution of the planet. Some theories suggest it is entirely a result of internal processes, such as a distinct difference in mantle convection underneath the two, while other theories postulate that the northern hemisphere was altered by a massive impact, or a collection of basin-forming impacts.

Tectonic Activity

Mars has experienced dramatic and widespread tectonic activity. In the past the crust was cracked repeatedly as a result of internal processes such as volcanism. Most famously, the massive canyon system known as Valles Marineris was created predominantly by tectonic processes, not water erosion as once was speculated. The Valles Marineris complex is the largest canyon complex in the solar system, stretching more than 4,000 km around Mars. It originates at one end near the volcanic zone known as the Tharsis Region. Although the exact method of formation for the canyon system is still being studied, it is speculated that initial cracking was the result of uplift in the volcanically active Tharsis area. Eventually, so much lava was extruded in this zone that the planetary crust was overloaded and further split under the strain. The canyon complex has certainly been modified by other processes since then, such as water flow, mass wasting, and wind erosion.

Today most of the breaking and shifting of the surface is quite minor and is the result of processes like landslides. However, marsquakes are to be expected, probably more often than one would expect moonquakes on the Moon. The weight of volcanic products in the Tharsis region is still not yet fully supported by the rocks in the area. This is a source of continual stress for the lithosphere in that area. Also, Mars is yet in its last stages of planetary cooling, and there may still be very minor levels of geologic activity that can contribute to small amounts of seismic activity.

Volcanic Structures

Volcanism was once the dominant process changing the surface of Mars. The evidence for this includes vast expanses of flood basalts and lava flows, as well as numerous volcanoes. Mars is home to the largest volcano in the solar system, Olympus Mons. Olympus Mons was known to astronomers as Nix Olympica ("Olympic Snows"), and was still visible during planet-wide dust storms, leading Schiaparelli to conclude it was a high mountain. The mountain is roughly 27 km high and over 500 km across, surrounded by a cliff up to six kilometers tall.

Olympus Mons is one of several volcanoes in close proximity to one another, an impressive volcanic structure known as the Tharsis Uplift or Tharsis Bulge. The origin of the Tharsis Bulge is likely related to the lack of plate tectonics on Mars; plate motion on the Earth might lead to a string of volcanoes over a hot spot, like the Hawaiian chain, but on Mars volcanism would have occurred in the same place for the entire duration of the hot spot.

Crater counts on some very small, localized flows suggest that Mars might have had a minor amount of volcanism still happening in the very recent past,

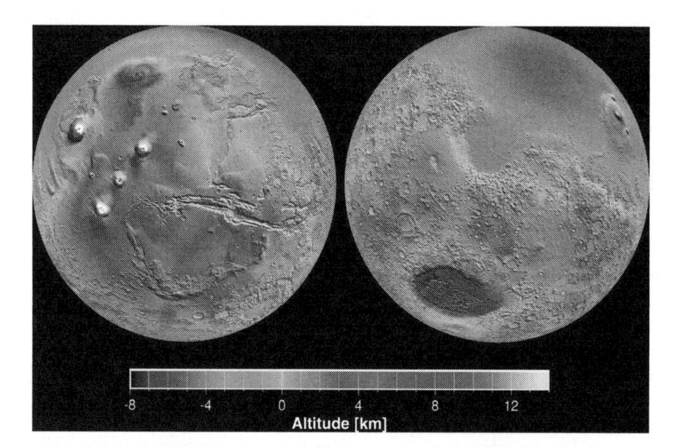

Figure 12.3 The laser altimeter carried by *Mars Global Surveyor*, called MOLA, compiled a very high quality topographic map of Mars. Easily seen is the overall difference in elevation between the northern and southern hemispheres, along with the large Hellas Basin in the southern hemisphere and the four volcanoes of the Tharsis Bulge.

only a few million years ago. There is further, if indirect evidence for recent volcanism in the form of methane in Mars's atmosphere. Because it is not stable, it must be of recent origin, consistent with recent volcanism. It is not impossible there is yet some tiny amount of activity going on today. Such places of current volcanism would be of great interest to planetary scientists, both because of the information they would provide about Mars and its processes, but also because they would be the best places to search for liquid water today.

Cratering Record

Cratering has been and continues to be an important process shaping the surface of Mars. The thin atmosphere helps to filter out some small impactors, but others find their way to the surface even today, and crater formation continues. Still, the atmosphere is more than substantial enough to allow wind erosion to take its toll on existing craters, and since it was thicker in the past, the oldest craters on the planet have been subject to water and ice erosion as well as wind. In spite of this, more than two dozen basins are identifiable, including the Argyre and the Syrtis Major basins, along with the massive Hellas basin measuring almost 2000 km in diameter.

Mars has some craters with a very distinct morphology. These are the "rampart" craters, or craters with splash-like ejecta blankets. They appear to have ejecta that was made of mud or some other substance that moved as a fluid while emplaced. The most likely theories for their formation include water, in the form of ice, being melted and mixed with the target material during the impact processes, but the Martian atmosphere may have played an important role as well. Investigations into the formation of these unique craters are necessary for the identification and characterization of water and subsurface ice in Mars's past and present.

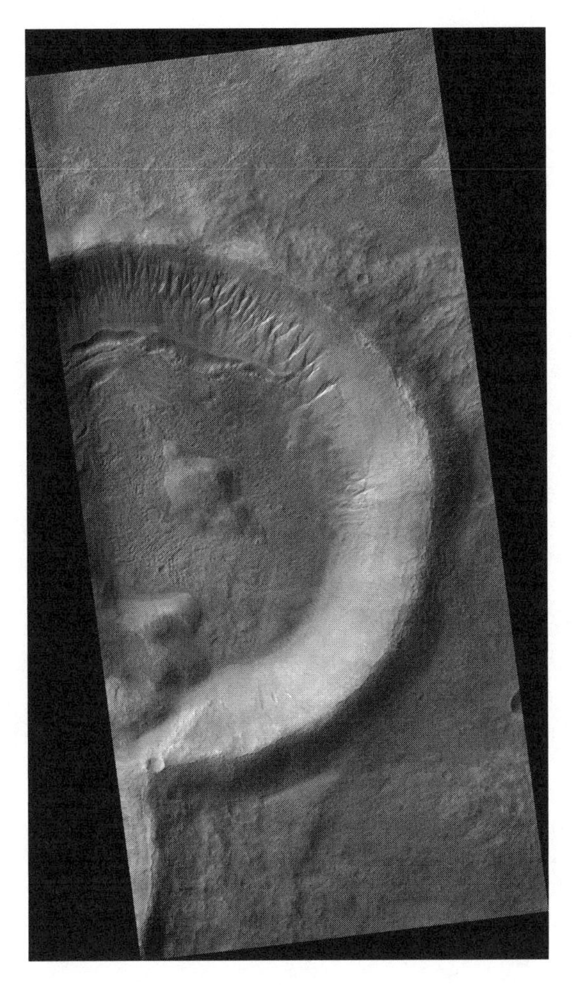

Figure 12.4 This crater, observed by the *Mars Reconnaissance Orbiter*, has ejecta that looks like flowing mud. This ejecta, near the top of the image, is evidence that ice or water was mixed in to the surface when the impact occurred.

Aeolian Activity

Aeolian processes are the wind-driven processes that shape specific features of a planet's surface. The wind can erode features, carry material away from an area (**deflation**), or leave sediment behind (**deposition**). The surface of Mars has been dominated for some time by the deflation of dust in some areas and the deposition of it into others.

The steep sides of canyon walls reveal layer after layer of dust, or in some places dust and lava flows. Massive dust storms on the planet are common, and sediment is moved from one place to another, building up layers in the new terrain and eroding the sediment from others. For instance, this can be seen in areas near Planum Boreum, which reveal an ancient landscape that has been hidden underneath and is now being exhumed by erosion. Crater counts indicate that this underlying terrain may have been blanketed for

Figure 12.5 On the bottom is a view of Victoria Crater, a 730 m crater in Meridiani Planum, taken by the *Mars Reconnaissance Orbiter.* Near the top edge of the crater is a thin set of lines: the tracks of the Mars Exploration Rover *Opportunity.* A view of the same crater from *Opportunity* is shown at top. The ability to observe the same area from orbit and in situ has provided great insights to Mars scientists.

millions of years. The existence of such formations indicates that Martian aeolian history is highly complex and will be very difficult to fully unravel. Surfaces might not really be the age suggested by crater counts if they have been "protected" by thick blankets of dust in the past.

Mars is host to a vast number of smaller scale aeolian features including wind streaks, ridges etched by the wind called yardangs, and ventifacts carved into the rocks. One of the most unique features created by wind are **dunes**. Frozen or **stabilized dunes** do not move—something like vegetation or ice has rooted them in place. But normally dunes are highly mobile. They move, shape, form, curve, pile up, and separate out in a fashion that seems almost lifelike. Dunes require a source of particles and persistent winds that blow in only one or two directions. Different kinds of dunes will form depending on the availability of particles and the nature of the winds.

But not just any particles will do to form dunes. Particles generally move across a surface in one of three ways. If the winds are higher and the particles are smaller, then the particles will travel relatively long distances, entrained in the winds by **suspension**. If winds are relatively low and particles are big, then

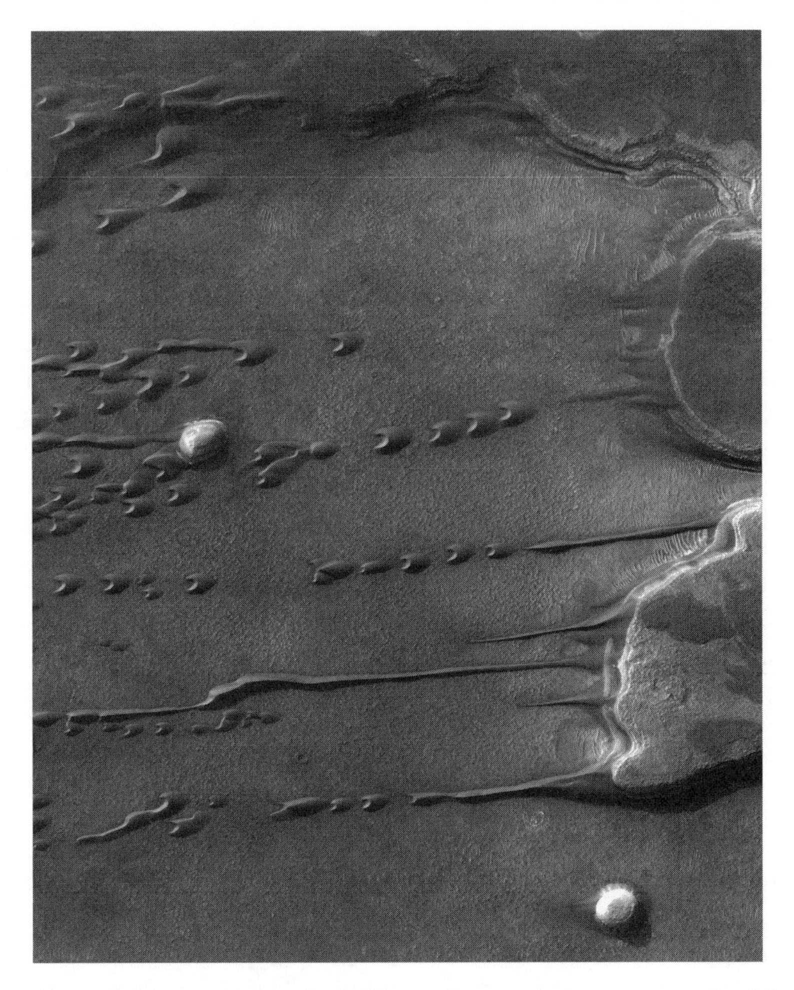

Figure 12.6 This image of the surface of Mars shows two flat mesas on the right. The regional wind is blowing from the right to the left, drawing particles out before it into long streaks and breaking up into arc shaped dunes called barchans. Barchans form with their "horns" facing downwind.

the winds can only push or roll the particles forward in nearly constant contact with the ground in a form of movement called **creep**.

In between creep and suspension is a particular form of particle motion called **saltation**. A saltating particle practically bounces across a surface. It is first launched by winds that move over it, creating a low pressure zone above the particle. This effect is called the **Bernoulli effect**, and is one of the reasons why airplanes can take off from the ground. Fast moving air over a plane's wing causes a low pressure zone above the wing, which puts an upward lift force on it. In the case of our particle, it is pulled up from the ground and temporarily launched into the wind flow in a trajectory shaped like a parabola. It is too heavy to suspend, however, and drops back down to the ground. When it impacts, it assists other particles in being launched up into the wind, and they also saltate, and continue the process.

Saltation seems to be the key to dune formation, since if particles and winds do not fall into the appropriate ranges for wind speed, atmospheric density, particle size, and more, then dunes will not form.

And yet they seem to form readily in any environment with fluid flow and available particles, including Mars, Earth, and Saturn's moon Titan. One of the reasons that dunes are so commonplace on the Earth is that the necessary size for a saltating particle in the usual pressures and wind speeds at the surface is the same size as a naturally forming particle of sand. Silicate sand has a characteristic size due to the strength of quartz. Much sand on our planet is predominantly made of quartz, which begins as rocks and is slowly eroded. When it reaches the size of a sand grain, the particle is very strong for its size and very hard to break or erode further. They can scour, erode, and bound along the surface and still survive. And so the abundant silica we have in our environment erodes into long-lived sand particles, and they are just right for saltating. The average size of a dune particle on Earth is about 250 µm, and particles must be at least 50 µm or so to saltate at all. Ralph Bagnold, a British engineer and soldier of the World War II era calculated the following equation for the mass of sand passing down a lane per unit time (q):

$$q = C\sqrt{((\rho/g)(d/D)u^{*3})}$$

where C is an empirically derived constant, ρ is the density of air, g is the gravitational acceleration, d/D is the ratio of the typical particle size to a reference size (250 μm in the original work) and u^* is the "friction velocity," measured by other means.

On Mars, the situation is more complex. The atmosphere is much less dense, and silica sand is less common. We can plug in the values for atmospheric density (roughly 9 percent that of Earth) and surface gravity (roughly 38 percent that of Earth) for Mars compared to Earth to find saltation conditions for Mars. When doing so, we find lowest possible size for saltation is roughly 210 μm, and so the average size of a Martian dune particle must be relatively large. The most abundant particles on the Martian surface are basalt, which are remaining bits from eroded lava flows. But small bits of basalt are not very strong, and will not saltate too many times before breaking into pieces so tiny they move from the saltation regime into the suspension regime. But theoretical issues aside, there must be abundant particles of the proper size on Mars, even if they are relatively weak, because the dunes are everywhere.

Water and Ice

There are several types of ancient water-cut valleys and river channels on Mars, the most impressive of which appear to have been carved by episodes

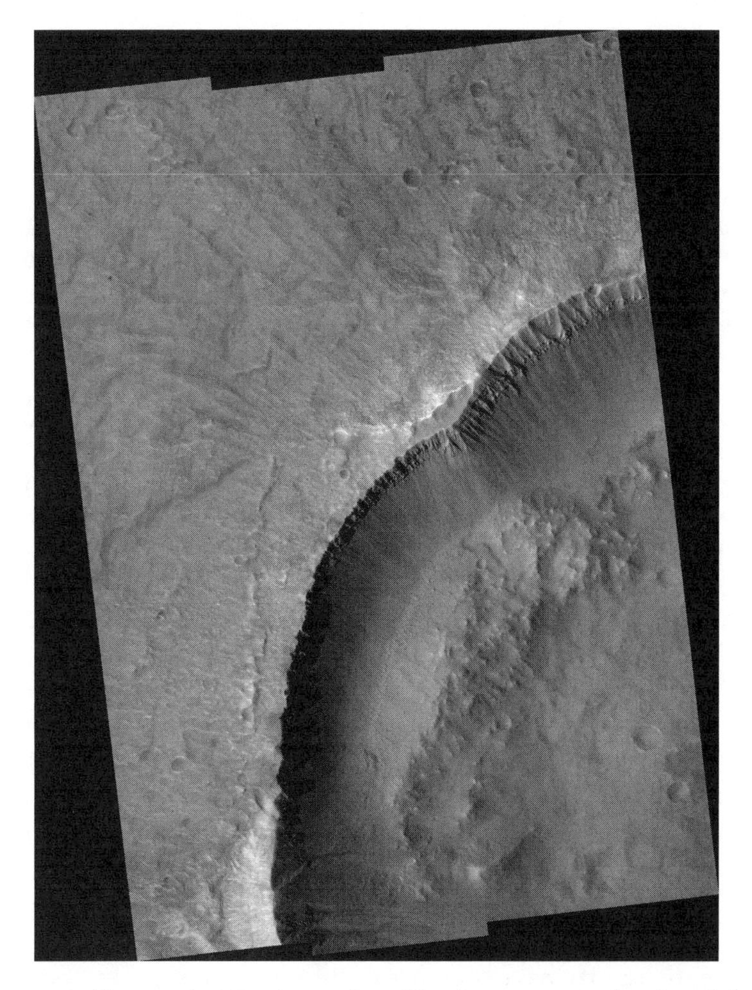

Figure 12.7 Gullies on the Martian surface, like those seen on the rim of the crater above, may be cut by water running from aquifers beneath the surface, or localized melt of subsurface ice. Or perhaps they are neither, and instead as some scientists postulate, they are landslides.

of catastrophic flooding. As noted, the planet has undergone different eras of heating and cooling of the surface related to its chaotic changes in planetary tilt, as well as from volcanic activity. The eras of greater heating would have caused large amounts of subsurface ice to melt, then as areas of the planet cooled (with the change in tilt or subsidence of volcanism) water would freeze once again. Some of the river valleys resemble networks of streams and tributaries such as one might find on Earth. They could be related to local regions of melt and runoff, or perhaps very localized rainfall in times of thick atmosphere. The smallest scale features are gullies found inside crater walls and in other places on the Martian surface. These seem to be caused by "springs" of water, formed by short-term melt of subsurface ice or areas tapping into small aquifers below ground. Some evidence exists to suggest that some of these regions are still active, being

Life on Mars

The search for life on Mars has taken some unexpected turns. The initial results from the *Viking* lander experiments were first interpreted as indicative of life before a rapid reanalysis concluded otherwise. The greatest controversy in this subject, however, occurred in the mid-1990s and centered on the meteorite ALH 84001. Structures inside the meteorite were interpreted as fossil bacteria, evidence of life on Mars. Naturally, this spurred great interest in the meteorite not only in the planetary science community but among biologists as well.

After the initial reports, additional analysis began to show a complicated story. Some scientists showed that the putative fossil structures could be generated by abiotic processes. Biologists argued that the tiny size of the fossils precluded any RNA, and thus any chance of life. Some argued that contamination from its residence time on the ice in Antarctica was the explanation. Others continued to argue that ALH 84001 contained evidence of Martian life.

As of this writing, most planetary scientists find the evidence for life in this meteorite unconvincing. Nevertheless, the search for life continues to drive some scientists to push for more Mars study and eventual return of samples that are freshly collected from the Martian surface, with no chance for contamination.

formed today. These gullies might be the only form of water found in a (briefly) liquid state on the planet, or they may simply be landslides, and not related to liquid water at all. Planetary scientists continue to study these features in an attempt to ascertain which formation mechanism is most likely and what it may imply for the existence of water on Mars.

Many of the large runoff channels empty into the lower northern plains. If there was sufficient water, and if the atmosphere was thick enough at the time, then there may have been oceans of standing water on the surface of ancient Mars. In fact, there may have been more than one era when oceans of various sizes existed. On a smaller scale, there are some dry rivers that empty into local lowlands or into the circular depressions formed by impact craters. In such places there may have been lakes of liquid water, and smooth floors of possible sediment in these areas provide some evidence to support this idea.

Mars has two polar caps. They are permanent within the current climate, but change in size each year with the seasons. Both are largely composed of water ice roughly two kilometers in thickness (though spread unevenly). As mentioned above, the polar caps also contain carbon dioxide: the northern cap accumulates roughly a meter of frozen carbon dioxide during the northern winter, while the southern cap has several meters of permanently frozen carbon dioxide over the water ice.

An interesting form of layered terrain can be found near each, but is more extensive around the southern polar cap. This polar layered terrain appears to be formed by cycles that include dust deposition, ice sublimation, and erosion. The north polar cap is also home to massive expanses of dune fields.

The polar regions of Mars have been the target for two NASA missions, with the unsuccessful *Mars Polar Lander* (MPL) preceding the successful *Phoenix* mission of 2008. *Phoenix* landed in a region with large polygonal cracks, thought to be due to seasonal changes in permafrost volume. Abundant near-surface ice was confirmed by the mission, which was previously predicted by the orbiting *Mars Odyssey* spacecraft.

MAGNETIC FIELD

Mars does not have a dipole magnetic field of global proportions, such as the Earth possesses. But there is some residual magnetism that remains within some localized areas of crustal rock, as discussed in more detail in Chapter 8. These areas sprout "umbrellas" of magnetism above them that stretch out past the edge of the atmosphere. There are dozens of these areas with relatively strong magnetic fields, and most are in the southern highlands region. Their location might be a result of the highlands rocks being older, and therefore retaining more of a memory of a former time when Mars may have had an active dynamo and global field. These umbrella regions may serve to protect the areas beneath them, both the atmosphere and the rocks, from the solar wind. Or the opposite may be true, at least in the case of the atmosphere. These regions may allow the solar wind to couple with the magnetic fields and tear off chunks of atmosphere all at once.

In spite of the lack of a global dipole magnetic field, aurorae have been observed on Mars, although not in visible wavelengths of light. The aurorae

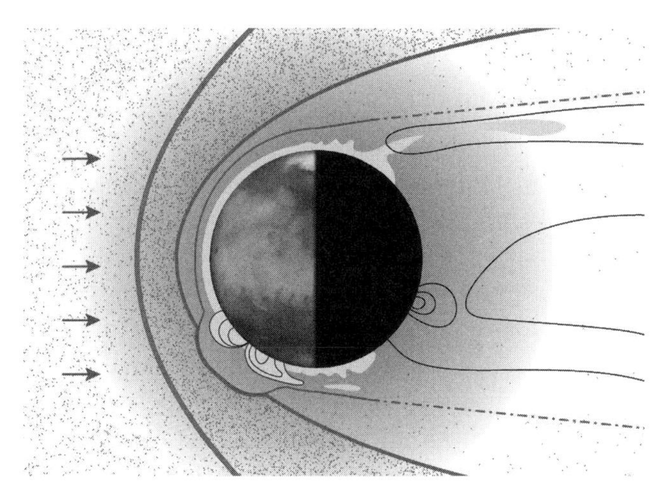

Figure 12.8 This diagram shows the planet Mars and a schematic representation of the "umbrella" shaped localized magnetic fields. The solar wind is coming from the left and blowing past Mars, an obstacle in its path. As it interacts with the localized magnetic fields, shown as arcs, it may remove bubbles or chunks of atmosphere and sweep them back away from the planet.

have been noted above areas with strong magnetic fields inherent in the rocks. This is consistent with the model of how aurorae form; they appear in places where a planet's magnetic field directs charged particles from the solar wind into the planetary atmosphere. On Earth, that is around the poles because of the presence of that planet's dipole field. But on Mars, all the magnetism is specific to certain areas of crust. The mystery of the formation of the aurorae relates to the problem of such relatively weak fields having the ability to accelerate the charged particles enough to produce these relatively high-energy phenomena.

CONCLUSION

Mars has fascinated scientists and nonscientists alike for centuries. It is the most Earth-like of the planets in many ways, and was for some time thought to harbor intelligent life, or be capable of it. The fact that it does not leads to questions of where and how the fates of our planets diverged. Mars may once have had a large ocean, but is today drier than the most parched desert on Earth. Mars has the most impressive volcanoes we know of, but today can only produce tiny amounts of activity, if any at all. These issues, as well as the relative accessibility of Mars will ensure that it remains an object of interest for scientists for many decades to come.

FOR MORE INFORMATION

The Mars exploration program is headquartered at http://mars.jpl.nasa.gov/ on the Internet. This site contains overviews of all of NASA's current and past missions. The European *Mars Express* site has its equivalent at http://www.esa.int/esaMI/Mars_Express/.

The Mars meteorite ALH 84001 is discussed in detail at http://www.lpi.usra.edu/lpi/meteorites/The_Meteorite.shtml.

An excellent collection of information about dunes on Mars is compiled at http://www.mars-dunes.org/index.php, including technical and nontechnical information.

Dr. William Hartmann, who has studied Mars for decades, wrote *A Traveler's Guide to Mars*, a tour of the planet with up-to-date science results woven in.

Jim Bell, member of the Science team for the Mars Exploration Rovers, has compiled over 100 3-D images from the *Spirit* and *Opportunity* rovers along with a discussion of the sights in *Mars 3-D: A Rover's Eye View of the Red Planet* (Sterling Publishing, 2008).

13

The Earth: Planet at Our Feet

INTRODUCTION

We often take for granted the fact that Earth is a planet. Exploring one of the most fascinating planets in the solar system is as easy as taking a good look around. It is in understanding the Earth that we have the best opportunity for understanding any other planetary body we come across. Like no other rocky world of which we are aware, the Earth possesses the full range of possible processes and features. The Earth has widespread oceans of liquid water on its surface, and of course the Earth has life, something that might be utterly unique in the universe, let alone our solar neighborhood. The planet retains a hot, active interior, which continues to drive a system of mobile crustal plates as well as frequent volcanism. On geologic timescales, the processes active on Earth, including interactions with the biosphere, serve to rapidly alter the planet's surface. The partially molten and rapidly rotating planetary core creates a strong global magnetic field. Our atmosphere creates constantly changing weather, the rains and winds of which continually sculpt the surface. With all of these amazing processes and phenomena right at hand, the Earth represents the best possible laboratory for the study of planetary science.

HISTORICAL BACKGROUND AND EXPLORATION

The Earth, ironically, was not recognized as a planet for most of human history. The planets of ancient times moved across the sky while the Earth remained unmoving. Ancient astronomers thought the Earth was at the center of the universe, but over the course of the last 500 years scientists have realized that far from being the center of everything there is, the Earth is merely the largest rocky planet circling an unremarkable star on the edge of a typical galaxy. Nevertheless, it obviously holds a special importance for humanity.

At roughly the same time as astronomers were concluding the Earth orbited the Sun, European explorers were beginning to get a true sense of our planet's size and scale. Technological advances in the 1700s allowed accurate latitudes and longitudes to be calculated, and accurate maps to be created. By the early 1900s virtually every part of the Earth had been mapped, whether the furthest reaches of the Arctic and Antarctic, to the jungles of South America and the islands of the Pacific. Detailed exploration of the ocean floor became easier in the late 1900s as bathyscapes and research submarines were built, with visits to the ocean floor and the discovery of black smokers and study of hydrothermal vents performed as the Soviet and American space programs were reaching toward the Moon.

The Space Age revolutionized our understanding of the Earth as well as other planets. The Earth's Van Allen radiation belts were discovered in 1958 by NASA's *Explorer 1* satellite (which was also the first successful launch by the United States). The first weather satellite was *TIROS 1* launched by NASA in 1960 and followed within a few years by the *Nimbus* program, which were the precursors of the weather satellites in use today. The first of the *Landsat* satellites were launched in 1972, and they continue to provide remote sensing observations of geology, agriculture, forests, and other earth science targets. Satellites have provided data that have allowed extremely accurate models of the Earth's atmosphere, surface, and interior. They have also provided the examples that have been followed in the

The Limits of Our Living

Human beings, like all species on Earth, use certain resources. For much of human history, it was thought that those resources could not be exhausted, either because the rate at which they were being used was slow, or for religious reasons. However, by the late 1800s it was realized that human society could in fact use enough of a resource to extinguish it altogether. This was first recognized with the extinction of species like the passenger pigeon or the dodo bird and the near-extinction of the American bison. Scientists have also concluded that the woolly mammoth was killed off by our Paleolithic ancestors.

Not only animal and plant species can be killed off by humans, however. While some resources do seem inexhaustible (we would be very hard pressed to use all of the oxygen in the atmosphere), there is concern that fresh water could be in short supply in coming years, and that the world's petroleum production could be in decline. Some see this as a further spur for expansion into space and colonizing other worlds. Many cultures and societies, however, are finding it a further impetus to conserve and lower consumption and find alternatives to scarce resources. With both secular and religious reasoning brought to bear, it is now more and more common for people to see humans as part of the Earth and its ecosystem, living in harmony with it, rather than simply a consumer in competition with every other species.

exploration of other planets, which are often based on the technology and operations pioneered by Earth-orbiting satellites.

PLANETARY INTERIOR

The Earth's interior is separated into three distinct chemical regions; the crust, mantle, and core. The crust is the thin outer shell of the planet, and primarily contains lower-density minerals including feldspar, quartz, and plagioclase. Minerals of moderate density, especially olivine, can be found in the thick mantle region. The core, which has an outer liquid region along with a solid center is composed of higher-density materials such as nickel and iron.

Along with distinct chemical regions, the Earth also has two distinct mechanical regions. In the **lithosphere**, which includes both the crust and part of the upper mantle, rocks behave in a predominantly brittle fashion. When stressed, the rocks of the lithosphere will most likely break. Below this is the **asthenosphere**, where higher pressures and temperatures allow for rocks to behave in a ductile fashion. When stressed, the rocks in the asthenosphere can warp, bend, and flow, even though they are in the solid phase.

The Earth is a relatively large rocky planet, and so has cooled more slowly than our other inner planet neighbors. Even so, it would have cooled completely by now if not for the existence of naturally radioactive elements in its interior. Heat from the interior drives convection of the mantle and powers the movement of the plates that make up the crust of our planet.

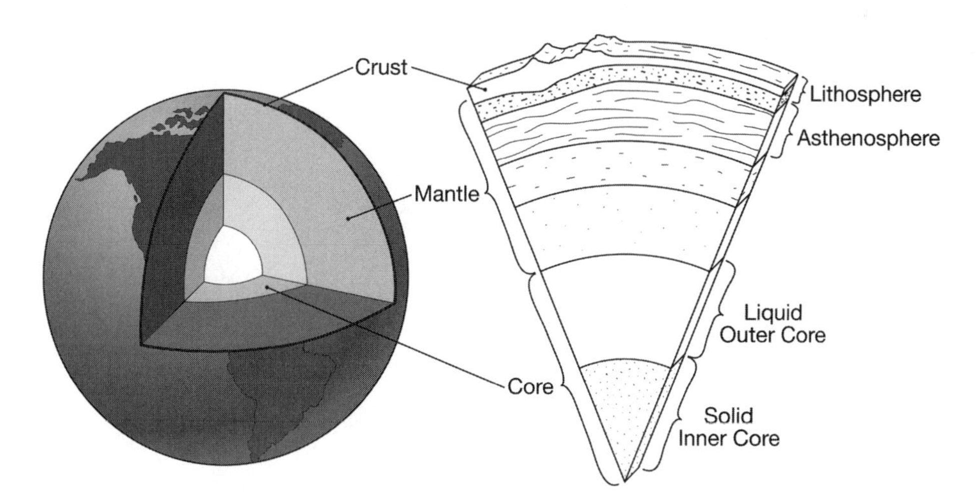

Figure 13.1 The Earth can be divided into layers based either on chemical composition or mechanical properties. The chemical divisions are core, mantle, and crust. The lithosphere is the region where rocks behave in a brittle fashion, found in the crust and very top of the mantle. In the asthenosphere, rock is more plastic and can bend, fold, and flow over long timescales.

ATMOSPHERE

The atmosphere of the Earth is composed of several layers of differing chemistry, pressure, and temperature. The bulk composition is largely nitrogen, with some oxygen, argon, and carbon dioxide. This was not the atmosphere the planet started with billions of years ago, its primary atmosphere. The Earth's original atmosphere was far more abundant in carbon dioxide, but plant life evolved to make use of that substance. Plants take in carbon dioxide and during metabolic processes separate the oxygen from the carbon. The carbon is incorporated into the plant itself, and the oxygen is released back into the environment. In addition to this process, much carbon dioxide eventually dissolved into our vast oceans. Much of our planet's carbon is now located largely in rocks like calcium carbonate, dissolved into our water, and making up plant biomass. In return, the plants

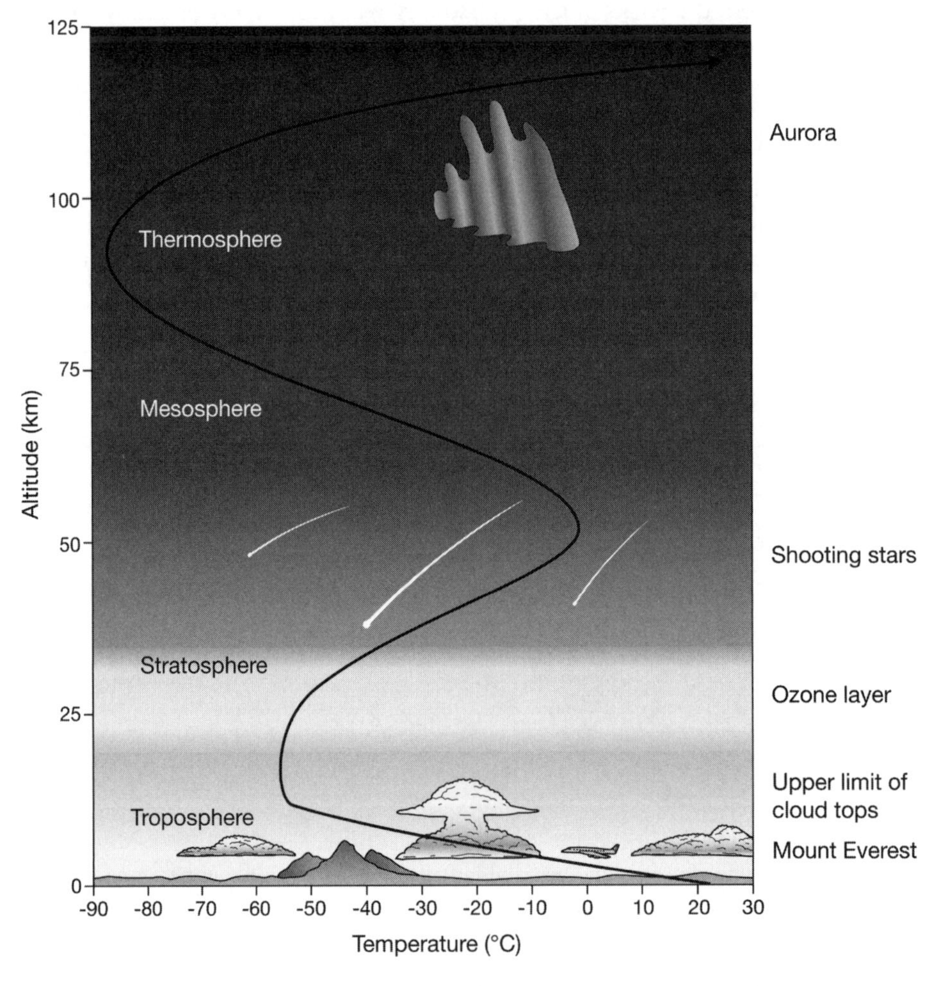

Figure 13.2 The temperature of the Earth's atmosphere does not simply decrease with height but instead has regions where it increases with height. Each time the temperature changes direction, a new section of atmosphere is defined.

The Ozone Hole

The layer of ozone (O_3) in the Earth's stratosphere absorbs most of the Sun's ultraviolet (UV) light and prevents it from reaching the ground. This is particularly important, since UV light can be harmful to life on Earth. From the 1930s through the 1950s, scientists studied the reactions that created and destroyed ozone in the stratosphere, including some that led to the destruction of much more ozone than might be expected. However, the molecules involved were all naturally found in the atmosphere, and therefore scientists expected the amount of ozone in the atmosphere was stable.

In the 1970s, further research found that substances called CFCs, commonly found in aerosol sprays, vastly increased the reactions that destroy ozone in the atmosphere. Because this material is *not* natural, and its use was increasing, there was thus a danger to the ozone layer. While the manufacturers of CFCs were dismayed at the theory, atmospheric scientists found overall depletions in atmospheric ozone compared to the era before CFCs, as well as vast regions over the poles that would seasonally lose most of their ozone (called "ozone holes"). Faced with the prospect of ozone losses over populated areas and significant increases in skin cancer as a result, the major CFC-manufacturing nations agreed to phase out CFCs and related chemicals in 1996. While the ozone holes continued to increase for several years after the ban, they have begun to shrink of late, taken as evidence the CFC ban was an effective remedy.

enriched our atmosphere with oxygen. Eventually, higher forms of life evolved to make use of this reactive substance, breathing it in, and exhaling carbon dioxide. Today, animals and plants both exist with a new atmosphere partially of their own making. They are intimately linked by a constant exchange of the atmospheric gasses each needs to survive—a highly symbiotic relationship.

Because of differential heating of the planet by the sun as well as the planet's rotation, our lower atmosphere has developed large-scale patterns of movement. Air circulates around the planet in cells like the near equatorial **Hadley cells**. These cells drive the major streams of air currents, such as the **jet stream**, across the United States. The movement of air in these cells, the evaporation of water, along with the heating of the atmosphere and planetary surface, all work together to create our planetary weather systems. The Earth has a huge variety of weather phenomena, including lightning, snow, rain, tornados, typhoons, and much more.

While any one individual weather phenomenon, like a storm, tornado, or snowfall is essentially random, general weather patterns can be predicted with the season and location. Larger averages in weather, precipitation, and temperature that span decades define a region's climate. But even climate changes over time. Our continents are always on the move. Ice caps grow and shrink, and ocean levels rise and fall. The amount of vegetative cover of land masses changes, as well. It is easy for us to imagine the climate we have now as the "right" one, or the "real" one, but this is simply the global climate with which we are comfortable and familiar. Because the Earth's global climate certainly will change in the future, the questions to ask are:

how much, in what way, when it will happen, and how will the change alter the conditions for life?

HYDROSPHERE

The Earth's abundant water can be found as vapor and water droplets in the atmosphere, as liquid rivers and oceans as well as ice and snow on the surface and filling pores and voids beneath the surface at the **water table**. It can also be found bound chemically or **adsorbed** into the rocks themselves. As geologic processes take place (like: evaporation, condensation, river downcutting, aqueous alteration of minerals, and the melting of rocks in subduction), our water cycles through all the systems and spheres of the planet. The complexity of this interaction means that on Earth water forms a "sphere" of its own, called the **hydrosphere**. Water exists on other planets, of course, but only on Earth does it move so freely between different systems.

SURFACE FEATURES

The Earth is home to a vast assortment of surface features including mountains, volcanoes, plains, ridges, graben, dunes, lava flows, canyons, impact craters, and meandering rivers. Many of the processes that create these features are working simultaneously—some enhancing one another, and others competing. Constant recycling of the Earth's surface means that no features remain from the planet's early history, and only a few isolated and particularly hardy mineral grains approach four billion years of age. Unlike the Moon, which still has craters that are billions of years old, most of the features on the Earth's surface have ages in the millions, or much less. So understanding the nature and evolution of Earth's surface features can be a challenging endeavor.

Plate Tectonics

All the inner planets show evidence of past tectonic processes, but the Earth possesses a unique and highly dynamic tectonic environment. The crust of the planet is separated into distinct plates, which move, and in some ways float upon the underlying mantle with speeds of a few centimeters per year. These plate movements provide a sort of planet-wide crustal recycling program. New crust is constantly being extruded along ridges in the ocean floor, while along **subduction zones** crust is being carried back down into the mantle and reworked.

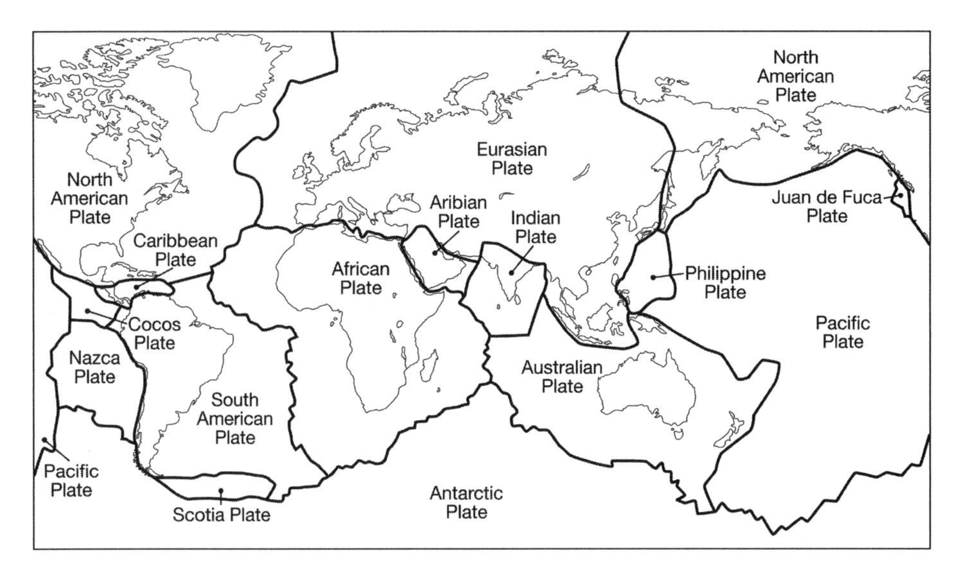

Figure 13.3 Image of global map with plate boundaries. The surface of the Earth is divided up into plates moving with respect to one another, some of which contain continental crust, some of which contain oceanic crust. Oceanic crust is denser than continental crust, and when they collide the oceanic crust is driven beneath the continental crust and destroyed. Oceanic crust is created where plates move apart, for instance where the North American and Eurasian plates are separating. Volcanoes are prevalent along the edges of plates, particularly the so-called "Ring of Fire" along the edges of the Pacific Plate.

At some plate boundaries, the plates slide past one another. Such regions can produce frequent earthquakes, as seen in the San Andreas fault zone. At other boundaries called **convergent boundaries**, plates run into one another. If the plates are both oceanic crust, one will be pushed below the other, and the result will be an underwater trench along the subduction zone and perhaps an arc of volcanic islands. If one of the plates is oceanic and the other is continental crust, then the oceanic plate, which is denser, will be pushed below the continent. A trench will form along the edge of the continent at the subduction point, and a line of volcanoes will emerge through the continental crust parallel with the rift. If both plates carry continental crust, then the plates will crumple together creating mountains in an **orogenic event**, or **orogeny**. The Appalachian mountains were originally formed this way, and the Himalayan mountains are being built in exactly this fashion today. **Divergent boundaries** are places where plates are moving apart. These generally form into the **rifts** where new crust is extruded. Most of these rifts are at the bottoms of oceans; ocean basins that the rifts essentially created by pushing apart the continental plates between them apart. But there are rift regions on the Earth that are currently found in continental settings such as the Rio Grande Rift and the African Rift Valley.

Volcanic Structures

Volcanism and volcanic features are found on Earth wherever magma can find its way to the surface. Usually, these areas are associated with plate boundaries or rift zones, but volcanism can also occur at "hot spots" or in places where the crust is experiencing some form of thinning or stretching.

The composite or stratovolcanoes of Earth are found predominantly at convergent plate boundaries where subduction is occurring. These volcanoes are steep-sided and experience explosive and dramatic eruptions. As one of the plates is subducted beneath the other, it carries trapped water down into the mantle. This lowers the melting point of mantle rock, which being less dense than the surrounding solid rock, rises. The magma eventually finds its way into a magma chamber, but in the process it becomes enriched in volatiles like water and carbon dioxide, and, it becomes more silicic from incorporating melted silica-rich crustal rock. By the time magma reaches a chamber, the pressure from the weight of the rock above it is substantially lower than it was when it was rising. This allows some of the volatiles to begin to come out of solution. This, in addition to other factors such as outside pressure on the chamber, as well as simple density contrast can drive the magma up toward the surface. The magma erupts explosively as the gasses come completely out of solution and force their way through the viscous silicic lava. Such eruptions are usually high in ash and other pyroclastic material. This is the reason composite volcanoes are steep. The highly viscous lavas and ash do not flow easily, and instead build up steep-sided structures.

Stratovolcanoes can produce highly dangerous "mudflows" called **lahars**. Such flows can appear to be made of mud, but are usually a mixture of water and pyroclastic material like ash. The water within a lahar can come either from the eruption itself, or from melted ice and snow from the surface near the volcano. Lahars can be scaldingly hot, fast-moving, thick and devastating, burying everything in their path with material that can have the consistency of quicksand or wet concrete. It is often this kind of flow that is the most hazardous feature of a volcanic eruption.

Shield volcanoes are comparatively benign, being composed largely of basaltic flows, which have a low viscosity. These produce a mountain structure with a very gentle slope, but one that can grow to be quite enormous. The Mauna Kea volcano in Hawaii is 33,000 feet high as measured from its base at the bottom of the ocean to the peak of the mountain. This is still dwarfed by Mars's Olympus Mons. The shield volcanoes of the Hawaiian islands are created by a "hot spot," a place where a plume of hot mantle material is rising up right under the oceanic crust. As the oceanic plate moves over the hot spot, it creates a line of volcanoes that have punched their way through the plate and built up over time. Eventually, the plate continues on, moves the current volcano off of the hot spot, and a new mountain is built behind the first. The line of undersea mountains trailing behind the Hawaiian islands is evidence for this phenomenon.

Hot spots can punch their way through continental crust as well as oceanic. There is a hot spot underneath the North American continent that appears to be moving to the northeast, but what is really happening is that the continent is moving westward beneath it. A trail of volcanism lies in its wake, with the oldest regions at the border of Oregon and Nevada radiometrically dated from 16 to 17 million years ago. The trail continues across Idaho, forming the Snake River plain, and leaving progressively younger volcanic regions behind. The hot spot now resides below the corner of Wyoming, with the last massive volcanic eruption, called the Lava Creek eruption, having taken place 640,000 years ago. The location is Yellowstone National Park, and the region is still very much volcanically active. In fact, the area is described as a "**supervolcano**," a volcanic region capable of generating some of the biggest volcanic eruptions on the planet. The Lava Creek eruption alone produced 1000 cubic kilometers of volcanic **tuff** material. For comparison, Mount St. Helens produced approximately four cubic kilometers of material.

In addition to ash, pulverized rock, lahars and such, volcanic regions produce lava flows. Some flows erupt from fissures and are so voluminous they blanket a vast area with volcanic rock. These are the "flood basalts" and they are found on Earth just as they are on other planets in the inner solar system. In the smaller flows that propagate from vents near shield volcanoes, one can see two distinct types of lava behavior. The first is **pahoehoe** lava, which flows more quickly and smoothly, and cools with a undulating ropy texture on its surface. The second is **a'a**, a type of lava flow that can look like a massive pile of moving black cinder blocks. When it cools, the surface does indeed look and sound like a broken and clinky pile of very rough glassy cinder blocks. Volcanism that takes place under water or under ice has certain unique features. One of those features is the production of **pillow lavas**, which are flows composed of pillow-shaped blobs of lava that cooled quickly upon being erupted into water.

Cratering Record

The Earth and the Moon share the same bombardment history, but the Earth's active surface processes have wiped out almost all of its impact craters. What remain are either those that formed very recently, or those that are so big they are hard to completely destroy. Scientists are still identifying impact craters on the Earth. This is because our instrumentation has improved, and our ideas of what to look for have changed. Smaller, recent craters such as Barringer (Meteor) crater in Arizona are easy to spot, but have been modified by wind, rain, and biologic activity. Such a crater on the Moon of the same size and age would still be nearly pristine. Most of these smaller craters have been accounted for, except those in very remote locations or those being confused with other structures such as lakes or salt domes. Most of the craters we are finding now are those that are almost impossible to identify by surface expression alone. These craters are being identified through a combination of several lines of evidence, such as finding characteristic impact products in the rocks, such as **shatter cones**, and seeing anomalies in global gravity data. More than 170 impact craters have been found so far, ranging in size from the tiny Carancas crater in Peru at 5 m across, to the huge Vredefort structure in South Africa, which is approximately 300 km in diameter. The oldest crater on Earth is probably the Suavjarvi crater, dated to approximately 2.4 billion years. The youngest craters are less than 1,000 years old, with Carancas crater forming in the twenty-first century.

Aeolian Activity

Like Mars, the Earth has abundant surface features created by wind-driven processes. The Earth possesses many active dune fields of all kinds, although not nearly as extensive as the biggest dune fields on Mars. Dust storms and sandstorms are common in arid regions, where in some cases dust and soil are being blown out of the area, and in other places sediment is being left behind. Wind is highly effective at erosion, and when it has dust or sand entrained within it, it becomes even more efficient, acting like a scouring agent. Aeolian activity is responsible for carving yardangs into the landscape, and on a smaller scale, is responsible for sculpting small rocks into wind-abraded ventifacts. When acting to deposit material, the wind can be responsible for dropping large quantities of sand in layers, changing depth and direction with the prevailing wind regime. Such sand beds can form into sandstone, and retain this **crossbedded** history of past wind regimes.

Water on the Surface

It is no surprise that with such ample water, the Earth is covered with features created when it flows across its surface. The most obvious of these are the lakes, seas, and oceans themselves, where large quantities of water pool into local topographic low regions. Lakes and seas are often places where sediment ends up, delivered by rivers, rain, or wind, and also eroded from the shoreline by waves, tides, and currents. Sediment sinks to the bottom of the body of water, and can later become **lithified** into **sedimentary rock**.

Water evaporates from the surface of oceans, lakes, and from the land. It also can sublimate from ice. All of these processes move water into the atmosphere, where it eventually precipitates out again in the form of rain, snow, etc. Rain that falls on mountains may end up percolating into the soil and rocks beneath and entering a local underground water aquifer, or it may run down the surface. Water that runs over the surface quickly

Figure 13.4 Wind-powered erosion and deposition can act on very large scales. This image from the MODIS satellite shows sand being blown off of the deserts of Africa en route to the Western Hemisphere.

funnels itself into small runoff channels that grow into creeks, then merge into larger and larger rivers as they flow downhill. All of these features, including channels and rivers are formed by water eroding the surface as it flows. Rivers may eventually cut very deeply into the land. This forms valleys, and eventually can form huge canyons, like the Grand Canyon in Arizona.

Rivers will cut downward until they reach a somewhat flat or open area. Assuming there is still an overall downhill gradient, and that water is not trapped in the area to form a lake, then the water will begin to erode laterally, instead of down. This creates classic river plains, where rivers bend sinuously back and forth in **meanders**. Around such rivers are flat regions that are nominally covered with water during times of flood. Flooding is a normal phenomenon that is a natural part of river systems. There are always times of greater or lesser rain or snow melt, usually seasonally, and flood plains are part of how the river system handles temporary increases in water. Flooding is only a problem in such areas when humans do not recognize the flood plain, and try to live there. Attempts to subsequently protect structures from flooding by erecting levees and such are not usually successful in the long term, since the river is self-designed to require the flood plain to properly move the volume of water and entrained sediment. Flood plains, like the area along the ancient Nile River, were chosen for farming precisely because they did flood seasonally. In the case of the Nile, new sediment would be deposited each year and refresh the farming conditions. Constructing levees or dams to protect structures ends up eliminating much of the reason that the region was of interest to people in the first place.

Ice is a common substance on the Earth's surface. Its formation, buildup, melting, and flow all contribute to shaping landforms. The largest reservoirs of ice on the Earth are the polar caps. They are not static, but change size as the global climate changes, raising or lowering sea levels in the process. Polar ice, which can be very long-lived, can also be a reservoir for older atmospheric gasses. The Earth's atmosphere has changed over time along with the climate, and gasses from the atmosphere become trapped inside polar ice as the years go by. Scientists can retrieve these gasses from ice cores and study the evolution of the Earth's atmosphere through relatively recent geologic history.

The Earth is also home to glaciers, which are formed by snow compacted into slow-moving ice flows. Glaciers can form at high latitudes or high altitudes in mountain ranges, wherever it is cold enough. Terrestrial glaciers have created a large suite of different surface features on the planet, including: smoothly scoured land, rounded and eroded hills and mountains, large boulders left behind in the form of glacial erratics, moraines of soil and rock, and small lake depressions called kettles.

MAGNETIC FIELD

As discussed in the chapter about magnetospheric processes, a large active dynamo can produce a planetary magnetic field. This is exactly the situation on Earth, where our rapidly rotating and partially liquid core powers an impressive global magnetic field. Our magnetic field operates as a shield, where charged particles from the sun are directed away from the surface of our planet, around to the sides and then back into the Earth's long **magnetotail**. Some charged particles are still channeled through to our polar regions, however, and interact with our upper atmosphere to produce colorful aurora.

The Earth's magnetic field can induce residual magnetism in rocks as they solidify from their initial molten state. The lavas extruded at mid-oceanic rift zones cool to form rocks that adopt a magnetic signature

Figure 13.5 In colder regions of the Earth, snow and ice can accumulate over years to form glaciers, which flow down valleys scraping rocks and soil ahead of them until they reach the sea or a climate where they melt and drop what they are carrying. Seen here is Columbia glacier in Alaska. The stripes inside the glacier are moraines, the areas of rock and soil scraped off of smaller glaciers that came together to form Columbia Glacier.

aligned with the Earth's magnetic field. These rocks are eventually pushed to either side as more lava is erupted and new rocks are formed. Oceanic crust is therefore formed in matching bands on both sides of the ridge, and is pushed further and further away in either direction as new rock is created in the middle. These bands preserve the signature of the Earth's magnetic field through recent geologic history. One of the main lines of evidence that suggested that global magnetic fields could spontaneously reverse their polarity (flip their magnetic north and south pole) was found in these bands of rock on the ocean floor. Over time, the rocks record several episodes of reversals of the Earth's magnetic field, and give some indication of the length of time between reversals, assuming a certain rate for the extrusion of the rock at the rift.

CONCLUSION

Our home planet is one of the most geologically diverse and interesting places in the solar system. Planetary scientists have been able to apply many of the lessons learned in studying the Earth to the other terrestrial planets, which in turn have provided insight that has been applied to this planet. Earth is a typical terrestrial planet in many ways, with an atmosphere intermediate in thickness between Venus and Mars, and a surface temperature intermediate between them as well. It has a magnetic field, as does Mercury, and it has had extensive volcanic activity, as have all of the bodies studied in this volume. However, it is also unusual in the composition of its atmosphere, the abundance liquid water found on its surface, and perhaps most obviously, the life that teems on its surface. Whether that life is unique in the solar system or not is still not fully answered, but the Earth is at the least, a very special place.

FOR MORE INFORMATION

A site with resources about plate tectonics, a critical factor in Earth geology can be found at http://www.ucmp.berkeley.edu/geology/tectonics.html.

The USGS (http://geology.usgs.gov/index.htm) has a set of pages focusing on a number of Earth Science topics including geology, magnetism, and hydrology.

The Environmental Protection Agency of the United States has compiled a number of links and resources related to possible climate change on the Earth (http://www.epa.gov/climatechange/). The British Broadcasting Company (BBC) also has a feature on climate change (http://www.bbc.co.uk/climate/).

http://www.spaceweather.com/ is a Web site specializing in interactions between the solar and terrestrial magnetic fields, although it also will carry other material. Aurora forecasts can be found here.

APPENDIX

Planetary Data Tables

The Inner Terrestrial Planets

Planet	Distance from Sun (AU)	Equatorial Diameter (km)	Mass (Earth = 1)	Density (g/cm^3)	Period of Rotation	Period of Revolution	Moons	Ring
Mercury	0.4	4,879	0.06	5.4	58.6 days	88.0 days	0	N
Venus	0.7	12,104	0.82	5.2	234 days*	224.7 days	0	N
Earth	1.0	12,756	1	5.5	23h56m	365.2 days	1	N
Mars	1.5	6,792	0.15	3.9	24h37m	686.9 days	2	N

*retrograde

The Outer Jovian Planets

Planet	Distance from Sun (AU)	Equatorial Diameter (km)	Mass (Earth = 1)	Density (g/cm^3)	Period of Rotation	Period of Revolution	Moons	Ring
Jupiter	5.2	142984	317.8	1.3	9h56m	11.9 years	63	Y
Saturn	9.6	120536	95.2	0.7	10h39m	29.4 years	60	Y
Uranus	19.2	51118	14.5	1.3	17h14m	84.0 years	27	Y
Neptune	30.0	49528	17.1	1.6	16h7m	164.8 years	13	Y

Glossary

A'a. A basaltic lava flow with a jagged, rough, and broken surface.
Accrete. To come together or collect, adding mass.
Acoustic Fluidization. The lowering of friction in a material through the input of seismic waves.
Adsorbed. A liquid, often water, accumulated onto or clinging to a surface, such as a sand grain.
Andesitic. A volcanic rock composed largely of plagioclase and feldspar.
Angle of Repose. The steepest angle at which a pile of loose material remains stable. This varies with the nature of the material and the water content.
Anorthosite. A rock composed of plagioclase feldspar.
Asthenosphere. The region of a planet below the lithosphere, where material will deform in a ductile fashion under stress, rather than break, or suffer brittle fracture.
Atmosphere. The envelope of volatile gasses, as well as free elements and ions, that surrounds a planetary body.
Attenuate. To lessen, decrease, or be removed.
Basaltic/Basalt. An extrusive, dark, igneous rock composed mostly of pyroxene and plagioclase feldspar (relatively rich in Fe and Mg elements).
Bernoulli Effect/Bernoulli Principle. The effect where fluid flow above a boundary (such as an airplane wing) is faster, and below is lower, causing an upward force along the boundary.
Bingham Fluid. A fluid that will only flow after a specific stress has been applied. The stress needed varies with the material.
Bingham Plastic. *See* Bingham Fluid
Body Wave. A seismic wave that travels within the interior of a body, instead of along the free surface.
Bow Shock. A region where a flow is suddenly decelerated; can refer to the area where the solar wind is decelerated from supersonic to subsonic speeds by interaction with the Earth's magnetic field.
Brittle. A material that will generally crack or break when stressed.
Calibrated. An instrument or device that has been checked against known standards so it can measure absolute, not just relative, changes or differences.
Capture Hypothesis. The hypothesis of the Moon's origin suggesting the Moon was not formed around the Earth, but was captured into Earth orbit.
Chart of the Nuclides. A variation of the periodic chart of the elements showing the isotopes of each element.
Cinder Cone. A hill composed of ash and cinders that forms as a cone around a volcanic vent or fissure.

Climate. Long-term trends in weather patterns, including average temperatures and amounts of precipitation.

Co-accretion Hypothesis. The hypothesis of the Moon's origin suggesting it formed in place around the Earth from the same, initial primitive material.

Comminuted. Broken up, crushed, smashed, pulverized.

Complex Craters. Craters of intermediate to large size that are characterized by flat floors, slumped crater walls, and central peaks or peak rings.

Compression. A stress regime where material is being compacted or pushed together.

Compression Wave. A form of wave where energy is transferred by the alternate compression and expansion of material.

Condense. The process of a gas changing state into a liquid as pressure increases or temperature decreases.

Conducting (electricity). A material that can carry an electric current.

Continuous Ejecta Blanket. The part of a crater's ejected material that falls within about one crater diameter of the crater rim, characterized by a continuous deposit of rounded debris and boulders.

Convection. The up-and-down cyclic movement of material as hotter packets of material move upward and expand, while colder packets move downward and contract.

Convection Cell. A particular region undergoing a circular pattern of convection.

Convergent Boundaries. Regions where two moving tectonic plates come together.

Coriolis Force. A force exerted on packets of atmosphere as a result of the Earth's rotation.

Coronal Mass Ejection. The sudden expulsion of mass from the corona of the Sun.

Creep. A form of sediment transport (along with suspension and saltation) where particles are pushed along the surface.

Crossbedded. A feature of some sandstones that shows alternating regimes of windflow and sedimentation that cause sand to be laid down in crosscutting layers.

Crystallized. Solidified and formed into a regular structured lattice.

Curie Point. The temperature above which a ferromagnetic material becomes nonmagnetic.

Curie Temperature. *See* Curie Point

D/H Ratio. Deuterium to Hydrogen ratio; a measure of the relative amounts of regular hydrogen atoms to those that contain a neutron in the nucleus.

Daughter Atoms. Atoms or elements produced by the radioactive decay of a larger atom.

Daughter Products. *See* Daughter Atoms

Decay. The act of an atom splitting into smaller particles that may be other atoms or pieces of atoms.

Deflation. A removal of sedimentary material or debris by wind or other fluid flow. The opposite of deposition.

Degassing. *See* Outgassing

Deposition. The emplacement of material, such as ash fall, crater ejecta, or lava flows. Usually refers to the emplacement of sediment out of suspension from a fluid or gas.

Deuterium. A hydrogen atom that contains a neutron in the nucleus along with a proton, instead of having only a proton, as in "typical" hydrogen.
Diamagnetic. A material whose electrons are aligned when placed in a magnetic field such that it is repelled by a magnet.
Differentiation. The chemical separation of materials within a planetary body, for example, the separation of metals toward the center of a planet.
Dipole. A material or object that has oppositely charged ends (positive and negative) called poles.
Discontinuous Ejecta. The part of a crater's ejecta blanket that breaks into a patchy deposit of debris, boulders, and secondary impact craters.
Dissolved. Made part of a solution.
Divergent Boundaries. A region at which two tectonic plates are moving away from one another, with new oceanic crust created at their boundary.
Domain. A region in a ferromagnetic material in which electron spins are aligned.
Ductile. A material that will generally flow rather than crack or break when stressed.
Dunes. Piles of particles (sand-sized on Earth) that move via saltation, in any medium.
Dynamo. The generation and maintainance of a magnetic field by the convection of a rotating ferromagnetic liquid.
Earthquake. The shaking of the ground resulting from the strain of released stress within the Earth.
Ejecta. The material thrown outward by the explosion that creates an impact crater.
Elastic. The behavior of a material under stress where deformation lasts only as long as the stress is applied. When the stress is removed, the material returns to its former shape.
Electric Current. The flow of electrons within a material under the influence of an electric field.
Electric Field. A region of space in which a charged particle will feel a force.
Electrically charged particle. A particle that exerts negative charge (an electron) or positive charge (a proton), or has an imbalance between its number of electrons and protons.
Electricity. A number of phenomena and processes, related by the involvement of charged particles.
Electromagnet. A magnet whose field is induced as the result of a changing electric field.
Electromagnetism. A force that acts upon charged particles, and that transfers energy by photons.
Epicenter. The point on a planet's surface that is directly above the source of a seismic event (earthquake, marsquake, etc.).
Equilibrium. A state of balance between forces or materials which will remain unless disturbed.
Erosion. The removal of material by various means including wind, water, landslides, glacial movement, etc.
Europium Anomaly. An enhancement or depletion of the element europium relative to other rare earth elements. A europium anomaly indicates plagioclase was crystallized from magma and either added or subtracted from the material in question.

Evaporate. The change of water from liquid to gas phase, usually outside of the context of boiling.
Exhumed. Revealed, exposed, or unburied.
Extension. A stress regime characterized by tension, a pulling apart of material.
Extruded. Material that has been pushed out from beneath a planetary surface onto that surface.
Failure. The sudden, discontinuous breaking of a material under stress.
Far Side. The hemisphere of the Moon that always points away from the Earth. Sometimes inaccurately referred to as the "dark side."
Fault. A fracture or group of fractures in rock or ice where strain (movement) has occurred.
Feldspar. A group of silicate minerals, including plagioclase, that are the most common minerals in the Earth's crust.
Felsic. Rocks rich in aluminum and silicon, with comparatively less magnesium and iron.
Ferromagnetic. Material whose electron spins align in the presence of a magnetic field, becoming magnetic themselves in the process.
Field. A space in which a particular force exists.
Fission Hypothesis. The hypothesis of the Moon's origin suggesting it formed from material that was spun off of the Earth when it was in a state of very rapid rotation.
Flows. Units of lava that have erupted onto a planetary surface.
Freeze. Changing state of a material from liquid to solid due to a lowering of temperature.
Frost Heave. A landform created by the repeated expansion and contraction of water as it undergoes repeated freeze-thaw cycles.
Gardening. Mixing of a regolith via impacts of various sizes.
Gas. A state of matter where molecules are not bound to one another in any structure.
Geiger Counter. An instrument used to detect the particles emitted by the radioactive decay of elements.
Geochemical. Referring to the chemistry of rocks and minerals.
Geomagnetic Storm. A period of intense aurora and disruption to the Earth's magnetic field, caused by solar flares.
Giant Impact Hypothesis. The hypothesis of the Moon's origin suggesting it formed, after the impact of a Mars-sized body into the Earth, from the coalescence of ejecta in Earth's orbit.
Glacier. An accumulation of snowfall that forms a large mass of mobile ice. This form persists from one winter to the next, and either flows under its own weight, or downhill due to the force of gravity.
Global Warming. A general upturn in the average global temperatures of a planet that may be due to random or episodic climatic variation, enhanced greenhouse effect, or human activity.
Graben. A trough or break in a solid surface created by tensional stresses.
Greenhouse Effect. Atmospheric warming caused by gasses in the atmosphere that are transparent to visible light but generally opaque to infrared. Heat re-radiated from the ground as infrared radiation is absorbed by these gasses, such as water vapor, methane, and carbon dioxide.

Hadley Cell. An idealized atmospheric convective cell where hot air rises near the equator, moves laterally away from the equator, sinks at mid-latitudes, and moves back to the equator near the surface.
Half-life. The time required for one half of a radioactive material to experience decay.
Heat. The flow or transfer of energy between systems at different temperatures.
Heavily cratered terrain. A terrain on Mercury postdating the intercrater plains.
Highlands. One of the main divisions on the Moon, characterized by mountainous terrain and often a plagioclase feldspar composition.
Hilly and Lineated Terrain. A terrain on Mercury characterized by hills and altered craters, antipodal to the Caloris basin and thought to be caused by that impact.
Hydrosphere. The system of liquid water, ice, and water vapor on the Earth (or another planet) and its circulation.
Hydrostatic Equilibrium. A state in which the force of gravity on a material is balanced by the strength of the material.
Hydrothermal System. A system of or pertaining to hot water and the interactions of hot water.
Ideal Gas. A gas that follows the ideal gas law, where pressure and temperature are linearly related to one another.
Impact Basin. A large impact structure, typically hundreds or thousands of kilometers across, and typically dating to very early solar system history.
Impact Crater. A structure, usually circular and of predictable depth and profile, resulting from the collision of an impactor into a target at supersonic speed.
Impact Event. A collision between a projectile and target, resulting in a crater.
Impactor. *See* Projectile
Intercrater Plains. A hilly terrain type typically found between large craters, suspected to be the oldest terrain on Mercury.
Interference Wave. A set of waves resulting from the interaction of two other waves, which amplify each other in some cases and cancel each other out in others.
Interplanetary Magnetic Field. The magnetic field generated by the Sun as found and measured throughout the solar system.
Inverted Stratigraphy. A sequence of flipped over stratigraphic rock layers formed as a consequence of the impact cratering process.
Ionized. Material, usually a gas, that has gained an electric charge via loss or gain of electrons compared to its neutral state.
Ionosphere. The highest part of an atmosphere, where interactions with solar radiation ionizes the molecules and atoms that are present.
Jet Stream. A relatively narrow current of air, which can have great influence on terrestrial weather.
Lahars. A flow of mud and pyroclastic material from an erupting volcano.
Late Heavy Bombardment. A period roughly 3.5–3.9 billion years ago of increased impact rate, before it diminished to its current value.
Lava. Molten rock (or ice in cryovolcanic systems) that reaches the planetary surface.
Lava Tubes. A tunnel formed within a lava flow when parts of a flow solidify, allowing hotter, still molten rock to run within the flow.
Leach. The removal of soluable minerals or other material from a rock by a solvent such as water.

Lithified. Material that has been transformed and hardened into rock.
Lithosphere. The region of a planet above the asthenosphere, where material will deform in a brittle fashion under stress, breaking rather than flowing.
Lobate Scarps. High cliffs on Mercury thought to be formed via compressional stresses, perhaps during planet-wide cooling.
Lodestone. A rock that is naturally magnetic, usually due to the presence of the mineral magnetite.
Longitudinal Wave. A wave in which the direction of wave oscillation is parallel to the direction of wave motion.
Love Wave. A transverse surface wave seen during earthquakes.
Lunar Highlands. The mountainous regions of the Moon. Predominantly composed of light-colored anorthositic rocks.
Lunar Mare. Low-lying, dark, relatively flat areas on the Moon that originated by basalt flows filling pre-existing impact basins.
Lunar Swirls. Regions on the lunar surface where the surface has been disturbed in an asymmetrical way. Currently thought to be associated with magnetic fields.
Mafic. An igneous rock that is relatively high in iron and magnesium, often containing minerals such as olivine and pyroxene.
Magma. Molten rock (or ice in cryovolcanic systems) that is flowing beneath the solid surface. When it reaches the surface it is called lava.
Magma Chamber. A reservoir below the solid surface containing magma.
Magma Ocean. A large, deep region of molten rock hypothesized to have existed on the Moon just after its formation, and possibly existing on other planets early in their histories as well.
Magnetic Field. A space in which magnetic force exists, and in which ferromagnitic materials align their spins.
Magnetic Moment. A measure of an object's net magnetic strength.
Magnetic Pole. The parts of a magnet attracted to or repelled by other magnets; like poles repel each other, opposite ones attract.
Magnetism. A force that attracts or repels materials depending on their electron spins.
Magnetosheath. The outer region of a magnetosphere.
Magnetosphere. The region of a planet where particle motions are primarily determined by magnetic forces.
Magnetotail. The portion of a magnetosphere opposite the solar wind, where the magnetosphere has an extended shape.
Mare/Maria. *See* Lunar Mare
Martian Crustal Dichotomy. The division of Mars into a high, heavily cratered southern region and a much lower, less cratered northern region.
Mascon. A gravity anomaly usually interpreted to be an area of increased density. A contraction of "mass concentration."
Mass Spectrometry. The study of an material's composition by breaking it up into its constituent elements, ionizing them, and measuring their charge to mass ratio.
Mass Wasting. The downslope movement of rock, ice, regolith, or other debris under the influence of gravity, possibly triggered by seismic activity, frost heaving, wind action, etc.
Meanders. River bends.

Melt. The liquid state of a material, or the act of a solid becoming a liquid.
Metamorphosed. Rocks that have been altered in mineralogy, chemistry, or texture by heating, pressurizing, or chemical processes.
Minerals. A naturally occurring material with a specific chemical composition and an ordered, usually crystalline, state.
Natural Log. A mathematical operation: the natural log of x is the power to which the irrational number *e* must be raised to equal x.
Near Side. The side of the Moon, that due to the Moon being tidally locked, always points toward the Earth.
New Moon. The phase of the Moon where the Moon's orbit takes it between the Earth and the Sun. In this geometry, no portion of the lit parts of the Moon can be seen from Earth.
Newtonian Fluid. A material for which a linear increase in stress results in a linear increase in strain.
Normal Fault. An inclined fault plane that suffers vertical movement, where the block below the plane has moved upward relative to the block above the plane. Associated with extensional stresses.
Normal Force. That component of a force perpendicular to a surface.
Ohmic Decay. The decay of a dipole magnetic field that is not being sustained.
Olivine. A silicate mineral that dominates the mantles of the terrestrial planets. Its most basic formula contains one silicon atom, four oxygen atoms, and either two atoms of iron or magnesium or one atom of each.
Orogenic Event. *See* Orogeny
Orogeny. A tectonic event responsible for intense folding, faulting, uplift, and thickening of the crust; responsible for building mountains under compressional stresses.
Outgassing/Outgas. The release of volatile gasses from a material.
P-wave/Primary Wave. The first (fastest) seismic waves generated from a seismic event like an earthquake; they travel by alternate compression and extension of material, either solid or liquid.
Pahoehoe. A basaltic lava flow with a relatively smooth, ropy surface.
Paramagnetic. A material whose electrons weakly align when in the presence of a magnetic field.
Parent/Parent Atom. A radioactive isotope that decays into daughter products.
Photodissociation. The breaking apart of a molecule by the input of a photon's energy.
Pillow Lava. Lava that is erupted underwater, resulting in a characteristic "pillow" shape. Often associated with mid-ocean ridge volcanism.
Plagioclase. A silicate mineral also including calcium and aluminum, one of the most common rock-forming minerals on the inner planets.
Planetesimals. Objects roughly one kilometer in size that accreted to form the planets early in solar system history.
Plasma. A state of matter where molecules are not bound to one another in any structure, as with gasses, but containing ions and capable of acting as an electrical conductor.
Plastic. Material that flows and deforms without breaking.
Plate Tectonics. A description of the division of the Earth's surface into regions, or plates, that move with respect to one another. This results in the creation and

destruction of oceanic crust and is one of the major processes in which heat is lost from the Earth.

Potassium-Argon System. The radiometric dating system based on the natural decay of potassium into argon, but using isotopes of argon to stand in as proxies.

Pressure. The force per unit area on a surface. It is also used as a measure of atmospheric thickness.

Primary Crater. A crater formed by the impact of a projectile, usually an asteroid or comet. Compare to secondary crater.

Projectile. An object that impacts a target, by convention the smaller of two objects involved in a collision.

Protoplanets. Objects of roughly 100 to 1,000 km in diameter that are in the process of growing to form full-fledged planets. Ceres and Vesta are thought by some to be leftover protoplanets in the asteroid belt.

Pyroclastic. Pertaining to volcanic materials ejected from a volcano, vent, or fissure; including fractured rock, gas, ash, and bits of lava.

Pyroxene. A common silicate mineral found in planetary crusts. Its most basic structure has two silicon atoms, six oxygen atoms, and two atoms of magnesium, iron, calcium, or a mixture of those elements.

Radiation. A type of energy transfer via emission of photons or particles.

Radioactive. Material that experiences radioactive decay; nuclei that change from one element to another by emitting gamma rays, electrons, or helium nuclei.

Radioactive Decay. *See* Decay

Radioactive Elements. Elements that experience radioactive decay.

Radionuclides. Atomic nuclei that are radioactive.

Rarefaction Wave. An area following a shock wave, where pressure is less than the initial pressure.

Rayleigh Wave. A type of surface wave associated with earthquakes.

Rays/Crater Rays. High-albedo (bright) features resulting from impact events; radially pointing away from the center of the crater and capable of extending many crater diameters away from the rim. Associated with very young, fresh craters.

Regolith. The layer of loose, unconsolidated material at the surface of a planet that can include broken up rock, dust, ash, glass, and other debris. Sometimes used synonymously with "soil" for planets other than Earth.

Remanence. *See* Remanent Magnetism

Remnant Magnetism. An imprinted magnetic field that remains detectable in a material after the imprinting field has been removed.

Resonance. A correlation of periods, where one period (for instance a rotation period) is forced by gravity to remain at a particular value related to another period.

Richter Scale. The scale of measurement in common use to describe the magnitude of a seismic event. Based on a logarithmic scale where a magnitude 3 is ten times stronger than a magnitude of 2.

Rifts. Long canyons or troughs often associated with volcanism and extensional stresses.

Rilles. Long valleys on the Moon (or other planets) thought to largely be related to lava tubes.

S-Wave/Secondary Wave. A type of seismic wave generated by a tectonic event; arrives after P-waves (primary waves), and moves energy via "up and down" motion of material (shear) and therefore does not travel through liquids.

Saltation. A type of sediment transport (along with suspension and creep) that moves particles by hops and bounces. Associated with a particular particle size depending on the gravity regime, nature of the atmosphere and density of the material.
Saturated. Can refer to a variety of limits or processes where a threshold is reached, and beyond which additional material or signal cannot be sensed or contained.
Scale Height. The distance above a planetary surface where the atmospheric pressure has fallen to 1/e (roughly 37 percent) of the surface atmospheric pressure.
Scarp. Cliff.
Scintillation. A flash of light caused by ionized material.
Secondary Crater. A crater formed by the impact of ejecta from another crater. Compare to primary crater.
Sediment. Any form of small debris, such as dust, pieces of rock, or other loose material that is carried, eroded, or deposited.
Sedimentary Rock. Rock composed of accumulated lithified sediment, such as lake sediment; often forms in layers.
Seismic. Relating to the formation of waves in solid/liquid bodies from tectonic events.
Seismic Waves. Waves produced from tectonic events such as earthquakes that carry energy through solid (sometimes liquid) material.
Shatter Cones. Features within rock outcrops that have suffered impacts, are often used as evidence an impact has occurred.
Shear Stress. A stress regime characterized by a forces that are not colinear.
Shield Volcanoes. Volcanoes built by repeated, overlapping lava flows, resulting in a shallow-sloped, broad structure.
Silicic. Material that is silica (SiO_2) rich.
Simple Craters. Bowl-shaped craters that are typically caused by smaller impactors, with exact size depending on the strength of a planet's gravity.
Simple to Complex Transition. The diameter at which simple craters transition into complex craters. Largely gravity dependant.
Slope Failure. The motion of material down a slope, often occurring when a slope becomes too steep beyond the angle of repose.
Smooth Plains. A major terrain unit on Mercury, composed of lava flows that were subsequently cratered.
SNC Meteorites. Meteorites of known Martian origin that include the Shergotty, Naklite, and Chassigny classes.
Solar Flare. A massive energy release from the Sun, accompanied by X-rays, gamma rays, and increased particle flux.
Solar Nebula. The original mass of gas and dust from which the Sun, planets, and the rest of the solar system originally formed.
Solar Wind. The constant stream of charged particles generated by the Sun.
Solute. The material in a solution that is mixed into the solvent. In salt water, salt is the solute.
Solution. A mixture of two or more substances, with at least one of the substances usually (but not required to be) liquid.
Solvent. The material in a solution into which the solute is mixed, typically a liquid. In salt water, water is the solvent.

Sputtering. Dislodging atoms from a material by impact with atoms or ions.
Stabilized Dunes. Dune formations that are no longer in motion due to some process such as lithification by salts, vegetation cover, etc.
Stable. An isotope that is not radioactive.
Standard. A benchmark to which other measurements are compared to ensure the consistency and reliability of data.
Strain. A change in a material's shape or volume as a result of stress; deformation or movement of the material due to stress.
Stratigraphic. Relating to rock layers and their characteristics such as their age, order of formation, size, and thickness.
Stratovolcano. A volcanic structure generally associated with more silicic, high viscocity lavas; forms structures of steep sides in layers of pyroclastic materials and lava flows.
Stress. A force that acts on a material such as a rock layer.
Subduction Zones. The zone where two plates meet, and one plate (oceanic crust) is forced down beneath a second plate (either oceanic or continental).
Sublimation. The change of a material from solid phase directly to gas phase, without a period spent as a liquid.
Superrotation. The rotation of an atmosphere (particularly Venus's) at speeds much greater than the planetary rotation period.
Supersonic. Faster than the speed of sound. The speed of sound varies depending on the material.
Supervolcano. A volcano capable of erupting more than 1,000 cubic kilometers of material, with long-lasting climatic and environmental effects. No supervolcanoes have erupted in historic times, though a handful are known.
Surface Wave. Seismic waves that travel along the surface (boundary interface) of a body instead of through the interior.
Suspension. A mechanism of transport for sediment in a gas or liquid. Particles in suspension can travel long distances before falling under the influence of gravity and contacting a surface.
Synchronous. A state for a satellite or planet in which its rotation period is equal to its period of revolution.
Target. The material into which a body or other projectile impacts to create an impact crater.
Tectonics. Processes that break, strain, and deform the crust of a planetary body.
Temperature. A measure of the energy in a material. In nature, energy will flow from hotter to colder materials until they reach the same temperature.
Tension. A stress regime where material is being pulled apart.
Terminal Lunar Cataclysm. A hypothesized short-term increase in the lunar impact rate roughly 3.8–3.9 billion years ago.
Tessera. A terrain on Venus thought to have been formed early in its history, characterized by intense faulting and deformation.
Thrust Fault. A type of fault in which two blocks in compressive stress move such that two surface points on opposite sides of the fault are brought closer together.
Transverse Wave. A wave in which the direction of wave oscillation is perpendicular to the direction of wave motion.
Triple Point. The pressure/temperature point at which all phases of a substance coexist; gas, liquid, and solid.

Tsunami. A massive wave produced in an ocean or other large body of water; usually created by underwater earthquakes, landslides, or volcanic eruptions.
Tuff. A rock formed from lithified pyroclastic materials.
Unstable. Refers to any phenomenon in a state liable to immediate change; also refers to isotopes that are naturally radioactive and will eventually decay.
Uplifted/Uplift. An area or region naturally pushed upward through bending of the crust.
Vacuum. A very low pressure, or a pressure relatively lower than another.
Viscosity. The property of a fluid that describes how much that fluid resists flow; higher viscosities are indicative of a greater resistance to flow.
Volatiles. Elements, compounds, or materials with low vaporization points.
Water Table. The zone beneath a planet's surface below which liquid water has filled available cracks and pore space.
Weather. Short-term changes in temperature, windspeed, and precipitation.
Weathering Product. A mineral created as a result of the alteration of another mineral via chemical reaction (often but not always involving water).
Wrinkle Ridge. Ridges in the lunar mare originating from the cooling and cracking of mare basalts.
Yield Stress. The threshold stress at which particular materials begin to strain, deform, or flow.

Bibliography

BOOKS

Beatty, J. Kelly, Carolyn Collins Petersen, and Andrew L. Chaikin, eds. *The New Solar System*. 4th ed. Cambridge: Cambridge University Press, 1998.

An excellent collection of articles about planetary science topics, including the inner planets.

Bell, Jim. *A Rover's Eye View of the Red Planet*. Sterling Publishing, 2008.

Bell, a member of the Science team for the Mars Exploration Rovers, has compiled over 100 3-D images from the Spirit and Opportunity rovers along with a discussion of the sights in Mars 3-D.

Bennett, Jeffrey, Megan Donahue, Nicholas Schneider, and Mark Voit.*The Cosmic Perspective*. 4th ed. San Francisco: Pearson Education Inc., Addison Wesley, 2007.

Binzel, Richard P., series ed. University of Arizona Space Science Series. Tucson: University of Arizona Press. 1979-.

The University of Arizona Press's long-running Space Science Series publishes cutting edge research on planetary sciences. This series is relatively technical, aimed at graduate school-level students and designed as a general reference for professionals. The series is currently 30 volumes in total, but the books in the series relevant to the topics discussed in this volume are listed below. The series is periodically updated, so, for example, *Venus II* supersedes *Venus*; however, earlier versions of books can still provide interesting and useful information and so are also listed below.

Bougher, Steven W., D. M. Hunter, and Roger J. Phillips. *Venus II: Geology, Geophysics, Atmosphere, and Solar Wind Environment*. 1998.

Canup, Robin M., and Kevin Righter. *Origin of the Earth and Moon*. 2000.

Chamberlain, Joseph W., and Donald M. Hunten. *Theory of Planetary Atmospheres An Introduction to Their Physics and Chemistry*. 2nd ed. Burlington, MA: Academic Press, 1987.

Chapman, Clark R., Mildred Shapley Matthews, and Faith Vilas. *Mercury*. 1989.

Colin, L., T. M. Donahue, and D. M. Hunter, with the assistance of Mildred Shapley Matthews. *Venus*. 1983.

Geodynamics. New York: Cambridge University Press, 2002.

This book is a technical introduction to tectonics designed for more advanced students.

Grinspoon, David Harry. *Venus Revealed: A New Look Below the Clouds of Our Mysterious Twin Planet*. Perseus Publishing, 1997.

This book takes stock of our understanding of Venus after *Magellan* in a popular-level work.

Guerrieri, Mary L., John S. Lewis, and Mildred Shapley Matthews. *Resources of Near Earth Space.* 1993.

Hartmann, William. *A Traveler's Guide to Mars.* Workman Publishing Company, 2003.

Dr. Hartmann, who has studied Mars for decades, wrote this tour of the planet with up-to-date science results woven in.

Jakowsky, Bruce M., Conway Snyder, Hugh H. Kieffer, and Mildred Shapley Matthews. *Mars.* 1993.

Matthews, Mildred Shapley, S. K. Atreya, and J. B. Pollack. *Origin and Evolution of Planetary and Satellite Atmospheres.* 1989.

Melosh, H. J. *Impact Cratering: A Geologic Process.* New York: Oxford University Press, 1989.

The classic technical consideration of impacts and impact processes is found here. Although out of print, those interested in details of the physics of impacts should find a library copy.

Plate Tectonics: How It Works by Cox and Hart. Palo Alto: Blackwell Scientific Publications, 1986.

This book is an excellent introduction to plate tectonics, a critically important process on Earth.

Strom, Robert G., and Ann L. Sprague. *Exploring Mercury: The Iron Planet.* New York: Springer, 2003.

This book provides a discussion of our state of knowledge on the eve of the *MESSENGER* mission.

Swihart, Thomas L. *Quantitative Astronomy.* Upper Saddle River, NJ: Prentice Hall Inc., 1992.

WEB SITES

The NASA Astronomy Picture of the Day often features images of the inner planets (as well as the rest of the universe!): http://antwrp.gsfc.nasa.gov/apod/astropix.html.

The Smithsonian and NASA provide a search engine for planetary science and astronomy technical papers at http://adsabs.harvard.edu/abstract_service.html, which can serve as a means of discovering in-depth information about the topics included here.

Google Mars (http://www.google.com/mars/) is a large database of visible, infrared, and elevation data mapped onto Mars. They have also done this for the Moon (http://www.google.com/moon/), and, famously, for the Earth (http://earth.google.com/).

Index

ABOUT THE AUTHORS

JENNIFER A. GRIER is a research scientist and education specialist affiliated with the Planetary Science Institute in Tucson, Arizona. She received her bachelor's degree in Astronomy and her Ph.D. in Planetary Sciences from the University of Arizona. Dr. Grier has published research focusing on impact crater chronologies, tectonic and volcanic histories of terrestrial planets, and thermal impact events recorded in meteorites, using data from spacecraft missions, telescopic observations, and laboratory experiments. She has also conducted educational research into student preconceptions about astronomy. Her instructional background includes presenting lecture and lab classes at the university level, online classes, and teacher education in science content and pedagogy.

ANDREW S. RIVKIN was born in New York in 1969. As a boy, the *Viking* and *Voyager* missions led him to an interest in astronomy that lasts to this day. A graduate of MIT with a doctorate in Planetary Sciences from the University of Arizona, Rivkin now works at the Johns Hopkins University/Applied Physics Laboratory, specializing in observing asteroids and analyzing their compositions. When not at the telescope or studying data, he's likely at a baseball game or listening to the Beatles.